Serious Managers Guide to AI Product Ownership

Understanding AI Products, Management, and the Unique Lifecycle

Jazper Carter

Cybersoft Publishing LLC

Fort Washington, MD 20744

Drclaude.net

First Edition March 2026

Table of Contents

Introduction 4

1 The AI Prod Owner Role: Accountabilities.........5

2 AI Prod Lifecycle: Stages, Handoffs,& KPIs....30

3 Reqs.: From Outcomes to Measurable Specs 60

4 Model Selection and Architecture Tradeoffs.....88

5 Data Strategy, Labeling, Feature Stores 111

6 Eval & Release Readiness for Gen. Sys 135

7 Human-in-the-Loop: Patterns that Scale 159

8 Monitor, Observe, and MLOps for PMs..........185

9 Release Mgmt, Sunsetting, Stewardship 213

10 Conclusion 240

Introduction

Most organizations didn't decide to become "AI organizations." It crept up on them. A pilot chatbot here, an analytics model there, a vendor tool with a recommendation engine under the hood. Then one day, a senior leader asks a simple question you're expected to answer: "Are we sure this thing is safe?" That question is why this book exists.

human-in-the-loop (HITL) means that human judgment is integrated into AI workflows at critical decision points. Humans do not simply observe the system; they actively review, validate, override, and improve it. HITL is essential for high-stakes use cases where errors have serious consequences, where context matters more than patterns, and where accountability demands that a human can explain and defend every decision. This chapter shows you how to design HITL mechanisms that protect safety and ethics without creating bottlenecks. You will learn about escalation thresholds, exception-review patterns, annotation feedback loops, and how to calibrate human oversight as systems scale. The goal is not to slow down AI, but to keep it aligned with mission and values even under pressure.

---

1 AI Product Owner Role: New Accountabilities

The call came on a Tuesday afternoon. Three separate engineering teams—fraud prevention, customer churn prediction, and supply-chain forecasting—had each deployed machine learning models into production over the previous eight months. None of them had coordinated. There were no shared data contracts, no common evaluation thresholds, no documented retraining schedules, and no agreed-upon owner for production incidents. When the compliance team began its quarterly audit, it found that two of the models were processing data categories that triggered regional privacy regulations, and the third had drifted so far from its original performance baseline that its outputs were no longer reliable—total estimated remediation cost: six weeks of engineering time and a formal regulatory disclosure.

The product manager responsible for the fraud-prevention model had been diligently tracking conversion rates and false-positive alerts. She had done everything a classic product manager would do. But she had never been told—and had never asked—about data lineage, model governance, or lifecycle accountability. That gap between what the role had been and what it now needed to be is exactly what this book addresses.

This chapter defines the AI Product Owner role in operational terms. It establishes the decision rights that distinguish this role from traditional product management, maps the path from business outcomes to model outcomes, describes the operating rhythms that keep AI products healthy, and provides the practical tools—a one-page charter, an escalation ladder, an onboarding checklist, and a metrics dashboard—that every AI Product Owner needs from day one.

1.1 Why It Matters

AI products do not fail the way conventional software products fail. A web application either works or it does not. An AI model degrades, drifts, hallucinates, and amplifies bias—often without a visible error message or a failed build pipeline. By the time a model's problems surface in a downstream business metric, weeks or months of harm may already have accumulated. The AI Product Owner exists precisely to prevent that silent failure mode.

Data stewardship is the first obligation. The AI Product Owner is responsible for knowing where training data comes from, whether it is fit for the intended use, and whether its use is authorized under applicable privacy and governance policies. This is not a data-science or legal responsibility in isolation—it belongs to the manager accountable for the product's outcomes. Responsible by design is not a slogan; it is a design constraint that must be embedded in the product brief before the first data pipeline is built.

Model stewardship is the second obligation. A model that passes evaluation today may fail silently next quarter as the real-world distribution of inputs shifts. The AI Product Owner is accountable for maintaining a retraining schedule, monitoring performance against agreed thresholds, and triggering intervention before degradation reaches users. Lifecycle accountability means owning the model from its prototype through eventual sunset—not just through its initial release.

The manager’s role in AI products is uniquely exposed. When a model causes harm—whether financial, reputational, or regulatory—someone must be identifiably responsible. That person is the AI Product Owner. This is not a burden to avoid; it is the source of the role’s authority. The clearest path to organizational clarity, mission-aligned outcomes, and stakeholder trust runs directly through an empowered, accountable AI Product Owner.

Diagram 1.1 – The AI Product Owner Accountability Map

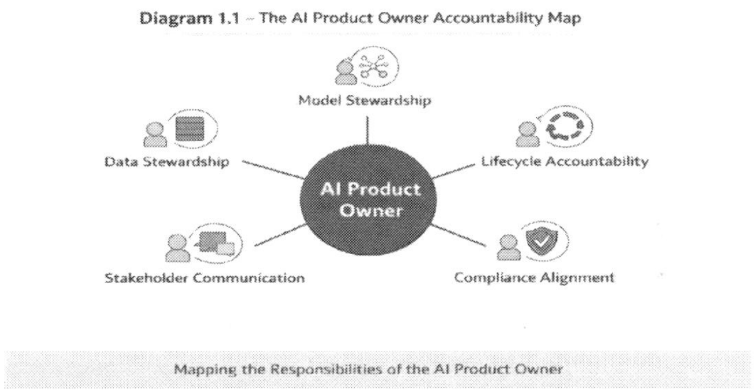

1.2 Defining the AI Product Owner Role

The AI Product Owner role is built on the foundation of classic product management—discovery, prioritization, roadmap governance, stakeholder alignment—but it extends that foundation in three directions that conventional PM frameworks do not address: model accountability, data governance participation, and technical risk ownership.

A traditional product manager owns features and user outcomes. An AI Product Owner owns all of that, plus the model that generates those outcomes. That expansion of scope is not incremental; it is categorical. The model is not a dependency to be handed off to an engineering team. It is a component with its own lifecycle, failure modes, and performance obligations. The AI Product Owner is the person who can be asked, at any point in time, "Is this model performing as intended?" and who is expected to answer with specificity.

1.2.1 What the Role Adds Beyond Classic PM

The additions fall into four operational domains:

- Model performance ownership: The AI Product Owner sets and defends model-level KPIs, reviews evaluation results, and makes go/no-go decisions for production readiness.
- Data accountability: The AI Product Owner approves data sources, reviews data quality reports, and escalates data governance issues to the appropriate stakeholders before they become model failures.
- Lifecycle management: The AI Product Owner owns the retraining schedule, monitors for data drift and concept drift, and initiates sunset planning when a model is no longer fit for purpose.
- Risk translation: The AI Product Owner translates technical model risks—bias, calibration error,

adversarial vulnerability, hallucination rate—into business-language risk assessments that executives and legal teams can act on.

These additions do not require the AI Product Owner to be a machine learning engineer. They require fluency—enough depth to ask the right questions, interpret the answers, and hold engineering and data science teams accountable for their commitments.

The AI Product Owner also plays a boundary-spanning function that no other role fills. Data scientists optimize models. Engineers build pipelines. Compliance officers assess regulatory exposure. Business stakeholders define success metrics. The AI Product Owner is the only person in the organization whose job description requires all four of those perspectives to converge into a single coherent product plan. That integration work—translating between technical constraints and business requirements, arbitrating among competing priorities, and maintaining a unified view of the product throughout its lifecycle—is the core value proposition of the role.

1.3 Decision Rights and the RACI Matrix

Ambiguity about who makes decisions is the single most common source of dysfunction in AI product teams. When a model fails in production, it should be immediately clear who owns the incident response. When a data source is proposed for use in training, it should be immediately clear who has the authority to approve or reject it. When a stakeholder requests a change to a model's decision boundary, it should be immediately clear whose sign-off is required. Without that clarity, decisions get made by default—by whoever moves fastest, speaks loudest, or has the most technical knowledge—rather than by the person with the right accountability.

The RACI matrix (Responsible, Accountable, Consulted, Informed) is the AI Product Owner's primary tool for establishing decision rights. It does not need to be elaborate. A one-page table covering the ten to fifteen decisions that recur most frequently in an AI product team is sufficient to eliminate most of the ambiguity that slows teams down.

1.3.1 Core Decision Categories

The decisions that require explicit assignment in an AI product context fall into five categories:

1.3.2 Product Strategy Decisions

These include: defining model-level KPIs, approving the product brief, setting the scope of the initial training dataset, and authorizing the pilot-to-production pathway. The AI Product Owner is accountable for all of these. Engineering leads and data scientists are consulted. Business stakeholders are informed.

1.3.3 Model Development Decisions

These include: selecting the model architecture, choosing evaluation benchmarks, setting training-data cutoffs, and determining acceptable performance thresholds. The data science lead is responsible for the technical execution. The AI Product Owner is accountable for the business relevance of the benchmarks and thresholds. Engineering provides input on infrastructure constraints. Compliance is consulted on the fairness and privacy implications.

1.3.4 Data Governance Decisions

These include: approving data sources for training and inference, reviewing data quality reports, authorizing the use of personal or sensitive data, and escalating data anomalies. The AI Product Owner is accountable. The data engineering team is responsible for execution. Legal and compliance are consulted. Senior business stakeholders are informed.

1.3.5 Release Decisions

These include: approving production readiness, authorizing staged rollout parameters, setting rollback thresholds, and signing off on release communication. The AI Product Owner is accountable for all release decisions. Engineering is responsible for execution. QA and security are consulted. Business stakeholders are informed.

1.3.6 Incident Response Decisions

These include: escalating production incidents, authorizing emergency rollbacks, communicating model failures to affected users or regulators, and initiating post-incident reviews. The AI Product Owner is accountable and must be reachable at all times. Engineering is responsible for technical remediation. Legal and communications are consulted for external-facing incidents. The executive team is informed.

Diagram 1.2 – AI Product Team RACI Matrix

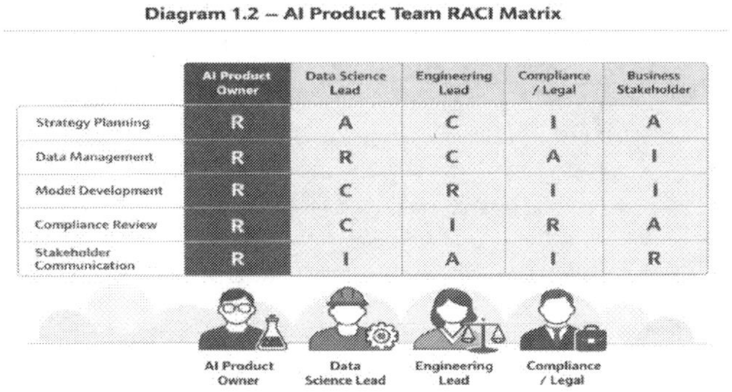

	AI Product Owner	Data Science Lead	Engineering Lead	Compliance / Legal	Business Stakeholder
Strategy Planning	R	A	C	I	A
Data Management	R	R	C	A	I
Model Development	R	C	R	I	I
Compliance Review	R	C	I	R	A
Stakeholder Communication	R	I	A	I	R

1.4 Translating Business Outcomes to Model Outcomes

The most frequent failure mode in AI product development is not technical; it is translational. A business stakeholder says, “We want to reduce customer churn.” An engineering team builds a churn-prediction model. The model achieves high accuracy on the test set. It ships to production. Six months later, churn has not changed—or it has gotten worse. The cause, almost invariably, is a failure to translate the business outcome into a model outcome with sufficient specificity.

The AI Product Owner is the translation layer. The role requires the ability to take a business objective—expressed as a revenue target, a customer satisfaction score, a risk reduction percentage, or a cost-efficiency goal—and decompose it into the specific model behaviors that would produce that outcome.

1.4.1 The Translation Framework

The translation process has four steps:

1. Anchor the business outcome. State the business objective in measurable terms: "Reduce voluntary churn among mid-market customers by 12 percent over the next fiscal year." This is the outcome that matters to the business. Everything downstream must trace back to it.
2. Identify the model's role in producing the outcome. The model does not reduce churn directly; it produces a prediction or a recommendation that enables a human action (a targeted retention offer, a proactive service call, a pricing adjustment) that may reduce churn. Be explicit about the causal chain: model output → human or system action → business outcome.
3. Define model-level KPIs that proxy for the business outcome. For a churn model, candidate KPIs might include: recall above 75 percent for high-risk churners, precision above 60 percent (to keep intervention costs manageable), prediction latency under 200 milliseconds for real-time use cases, and a fairness threshold to ensure that churn risk predictions do not systematically differ by customer demographic segment.
4. Set monitoring thresholds that trigger review. Each model-level KPI needs a production threshold: the value below which the AI Product Owner initiates a review cycle. Thresholds should be set before launch and documented in the model card.

This translation work is not a one-time task. Business objectives shift. Market conditions change. User behavior evolves. The AI Product Owner is responsible for reviewing the business-to-model outcome mapping quarterly and updating model-level KPIs when the mapping has drifted.

The vignette that follows illustrates what happens when this translation is done well.

1.4.2 Vignette: Preventing Costly Misalignment

A mid-sized financial services company was preparing to launch a loan-underwriting model to accelerate decisions on small-business credit applications. The engineering team had been tasked with improving approval throughput—the number of applications processed per day—and had optimized accordingly. The model achieved outstanding throughput scores on the validation set.

The AI Product Owner, newly assigned to the project two weeks before the planned launch, reviewed the evaluation artifacts and immediately flagged a gap. The model had been optimized for throughput but had never been evaluated against the default rate by credit tier—the business outcome the CFO had identified as the product's success criterion. When the team ran the missing evaluation, the results showed that the model’s approval patterns were biased toward applications that were easy to process, not toward those likely to repay. Projected default rates in the first six months exceeded acceptable limits by a factor of nearly two.

The AI Product Owner held the launch and negotiated a staged rollout: a shadow-mode deployment over thirty days in which the model's decisions were logged but not acted upon, followed by a narrow pilot covering five percent of applications in a single geographic region, with a weekly review gate before each incremental expansion. The shadow-mode data revealed two additional miscalibrations that required retraining before the model was production-ready.

The delay cost the company four weeks of launch time. The alternative—discovering the default-rate problem after full deployment—would have cost the company materially more in loan losses, regulatory scrutiny, and reputational damage. The AI Product Owner's workflow-level impact in this case was not a feature decision; it was a mission-aligned act of governance.

1.5 Operating Rhythms for AI Products

AI products require a governance cadence that does not exist in most conventional product management frameworks. A weekly sprint review catches feature bugs. It does not catch model drift, data quality degradation, or the gradual erosion of a model's fairness properties. The AI Product Owner must establish and sustain a set of operating rhythms tailored to the challenges of AI product stewardship.

1.5.1 Weekly Model Review

The weekly model review is a standing meeting—thirty to forty-five minutes—that brings together the AI Product

Owner, the data science lead, and the engineering lead responsible for the model's infrastructure. The agenda is fixed:

- Review model performance metrics against production thresholds from the previous seven days.
- Review data quality reports for any anomalies in training or inference data pipelines.
- Review the incident log for any alerts or escalations triggered since the last review.
- Review retraining schedule: Is the next scheduled retraining on track, and has the trigger condition been met?
- Decision: hold current production state, initiate investigation, or escalate to monthly review.

The weekly model review is not a deep-dive session. Its purpose is operational clarity: confirming that the model is behaving as expected, identifying early warning signals, and maintaining a documented record of review decisions. The AI Product Owner chairs the meeting and owns the decision log.

1.5.2 Monthly KPI Review

The monthly KPI review is a broader session—sixty to ninety minutes—that connects model performance to business outcomes. It includes the AI Product Owner, engineering and data science leads, the business owner or product director, and optionally a compliance representative. The agenda covers:

- Business outcome metrics: Has the model's production impact moved the target KPIs in the intended direction?
- Model health metrics: Accuracy, calibration, fairness, latency, and cost trends over the past thirty days.
- Data health metrics: Pipeline reliability, data freshness, and any changes in input feature distributions.
- Roadmap alignment: Are planned model improvements still aligned with current business priorities?
- Risk register update: Have any new risks materialized, and have previously identified risks been mitigated?

The monthly KPI review is also the forum for making roadmap decisions. If a model's performance is trending toward a threshold breach, the monthly review is where the AI Product Owner presents options: emergency retraining, rollback to a prior version, or a controlled reduction in the model's scope while a fix is developed. These decisions require business context and should not be made in a technical meeting room without business representation.

1.5.3 Incident Postmortems

Every production incident involving an AI model—whether it triggers a customer-facing failure, a threshold breach, or an unexpected behavior pattern—requires a written postmortem. The postmortem is the AI Product

Owner's primary accountability document. It records what happened, why it happened, what was done in response, and what changes have been made to prevent recurrence.

A postmortem that is completed within five business days of incident resolution and shared with the relevant stakeholders is a sign of a well-governed AI product. A backlog of unwritten postmortems is a sign that the operating rhythm has broken down.

The AI Product Owner owns the postmortem process. This does not mean writing every technical section personally; it means ensuring that the document is produced on time, that root causes are identified with specificity, and that action items are assigned with owners and due dates. Postmortems should be stored in a shared repository and reviewed quarterly for systemic patterns.

Diagram 1.3 – AI Product Operating Rhythm Calendar

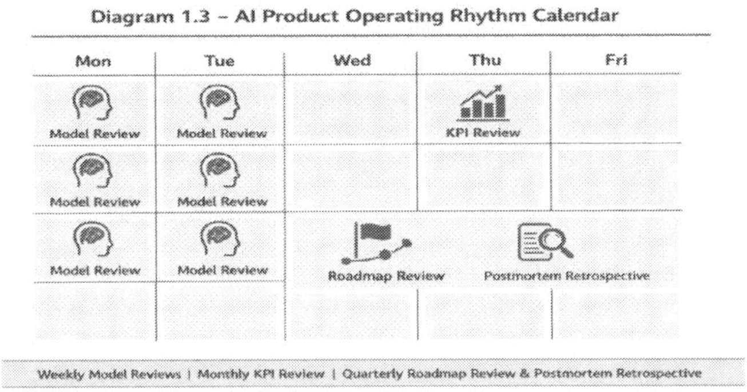

1.6 The AI Product Owner Charter

A product charter is a one-page document that establishes the AI Product Owner's mandate, accountability boundaries, and operating authority. It is not a strategic vision document. It is a governance document—a record of what the AI Product Owner is responsible for, who has authorized that responsibility, and what decisions fall within their unilateral authority versus those that require escalation.

Every AI Product Owner should have a charter in place before the first model enters production. The charter should be signed by the AI Product Owner and by the senior leader to whom they report. It should be reviewed annually and updated whenever the product's scope or the organization's AI governance framework changes.

The following checklist constitutes a minimal charter template:

- Product identity: Name of the AI product or model family, version, and production deployment date.
- Business objective: One-sentence statement of the business outcome the product is intended to produce, with a measurable success criterion.
- Model-level KPIs: The two to four model performance metrics against which the product will be evaluated, with production thresholds and review-trigger thresholds for each.
- Data governance summary: Primary training and inference data sources, data owner contacts, and any applicable privacy or regulatory constraints.
- Team and decision rights: Names and roles of the data science lead, engineering lead, compliance contact, and business owner, with brief RACI annotations for key decision categories.

- Operating rhythm commitments: Weekly model review cadence, monthly KPI review cadence, and postmortem commitment (within five business days of incident resolution).
- Escalation contacts: Names and response-time commitments for each level of the escalation ladder.
- Audit and review schedule: Date of the next scheduled model audit, data governance review, and charter renewal.

The charter serves a second purpose beyond governance: it is a communication tool. When a new executive joins the organization, a compliance audit is initiated, or an incident triggers a senior leadership review, the AI Product Owner can hand over the charter as the primary reference document. A well-maintained charter is evidence of a well-governed product.

1.7 Escalation Ladder for AI Issues

AI product issues do not always escalate on the same timeline as conventional software incidents. A model accuracy degradation may be tolerable for a few days while data science investigates; a model generating outputs that violate regulatory requirements must trigger an escalation within hours. The AI Product Owner must have a clear, pre-agreed escalation ladder that specifies who gets notified, at what threshold, and within what timeframe.

1.7.1 Level One: Technical Team Resolution

Threshold: Performance metric falls below review-trigger threshold but remains above a critical threshold; data quality anomaly detected but not yet affecting model outputs; latency spike within acceptable bounds. Owner: AI Product Owner and data science lead. Response time: within one business day. Actions: Investigate root cause, assess whether the issue is transient or persistent, and document findings in the incident log.

1.7.2 Level Two: Cross-Functional Review

Threshold: Performance metric breaches the critical threshold; data quality issue is affecting model outputs; model behavior appears inconsistent with the charter's defined guardrails. Owner: AI Product Owner escalates to the engineering lead, compliance contact, and business owner. Response time: within four business hours. Actions: Convene emergency review, assess rollback options, determine whether user-facing communication is required, and document the decision rationale.

1.7.3 Level Three: Senior Leadership Notification

Threshold: Model failure is causing material user impact, creating regulatory exposure, or generating adverse media risk. Owner: AI Product Owner escalates to senior product leader and legal/communications team. Response time: within one business hour. Actions: Initiate rollback if not already underway, prepare executive briefing (one page: what happened, what is being done, estimated resolution time), assess regulatory notification obligations, and assign a dedicated incident commander.

1.7.4 Level Four: Executive and Regulatory Escalation

Threshold: Incident has triggered a regulatory notification obligation, is generating material reputational risk, or involves potential harm to users. Owner: Senior legal or compliance leader, with the AI Product Owner supporting. Actions: Engage external legal counsel if required, prepare regulatory notification, implement communications plan, and conduct a full postmortem within ten business days of resolution.

Diagram 1.4 – AI Product Escalation Ladder

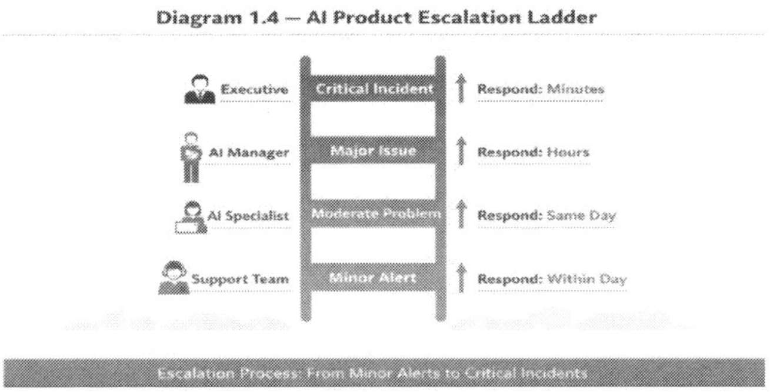

1.8 Onboarding Checklist for New AI Product Owners

A new AI Product Owner inheriting an existing model or product line must move quickly to establish situational awareness and continuity of governance. The following five-item checklist covers the minimum actions required in the first thirty days:

1. **Audit the existing model inventory.** Obtain a complete list of all models in production under your ownership. For each model, confirm: current production version, deployment date, last retraining date, owner of the training data, and current performance against production thresholds. If a model card or charter does not exist, create one immediately.
2. **Review the data governance posture.** Identify all data sources used in training and inference. Confirm that data use agreements are current, that privacy impact assessments have been completed for any sensitive data categories, and that data quality monitoring is in place. Flag any gaps to the compliance team within the first week.
3. **Establish the operating rhythm.** Schedule the first weekly model review and monthly KPI review before the end of your first week. Introduce yourself to the data science lead, engineering lead, and compliance contact. Confirm that each team member understands their role in the review cadence and has access to the relevant dashboards and documentation.
4. **Map the decision rights.** Review or create the RACI matrix for the product. Identify any decisions that are currently being made without a clear owner and escalate them for resolution. Confirm that you have the authority—granted in writing by your senior

leader—to make the decisions assigned to the AI Product Owner role.

5. **Validate the business-to-model outcome mapping.** Meet with the business owner to confirm that the model's current KPIs are still aligned with the business objective. Review any changes to the business context—new competitors, regulatory changes, user behavior shifts—that may have occurred since the model was originally defined. Update the model-level KPIs if the mapping has drifted.

1.9 Five Metrics Every AI Product Owner Must Track

The AI Product Owner is not a data scientist, but they must be fluent in model performance data. The following five metrics constitute the minimum viable monitoring dashboard for any AI product. They are not a complete observability solution; they are the indicators that allow a non-technical owner to maintain a credible, mission-aligned view of product health.

1.9.1 1. Business Impact Metric

This is the KPI that connects the model to the business outcome: churn rate, approval throughput, fraud loss rate, customer satisfaction score, or whatever metric the product is intended to move. The AI Product Owner should be able to state, at any time, the current value of this metric and whether it is trending in the intended direction. A model that is technically healthy but failing to move the business impact metric is not a successful AI product.

1.9.2 2. Model Performance Metric

This is the primary evaluation metric used during model development: accuracy, F1 score, precision/recall, AUROC, or mean absolute error, depending on the task. The AI Product Owner must know the threshold below which the model is no longer fit for purpose, and must ensure that this threshold is being monitored in production. When the metric falls below the review-trigger threshold, the weekly model review must produce a documented response.

1.9.3 3. Data Freshness and Quality Score

AI models are only as good as the data they consume. The AI Product Owner must track whether the model's inference inputs arrive on schedule, whether data quality checks pass, and whether the distribution of input features has shifted relative to the training distribution. A deteriorating data freshness or quality score is often the earliest warning sign of an impending decline in model performance.

1.9.4 4. Latency and Throughput

Users and downstream systems have latency expectations. A recommendation model that was designed to respond in under 100 milliseconds cannot degrade to 800 milliseconds without causing workflow-level impact. The AI Product Owner must track production latency percentiles (p50, p95, p99) and throughput against the service-level commitments made at launch. Latency degradation is frequently a signal of infrastructure issues, data pipeline problems, or increases in model complexity that have not been accompanied by infrastructure scaling.

1.9.5 5. Fairness and Disparate Impact Indicator

For any model whose outputs affect people—credit decisions, hiring recommendations, healthcare triage, content moderation—the AI Product Owner must track whether the model's performance varies systematically across demographic groups, user segments, or geographic regions. A fairness indicator does not need to be elaborate: a simple comparison of model performance across the two or three most relevant subgroups is sufficient to capture the most common disparate-impact patterns. This metric is not optional. It is a responsible-by-design requirement, and its absence constitutes a governance failure in itself.

Diagram 1.5 – AI Product Owner Metrics Dashboard

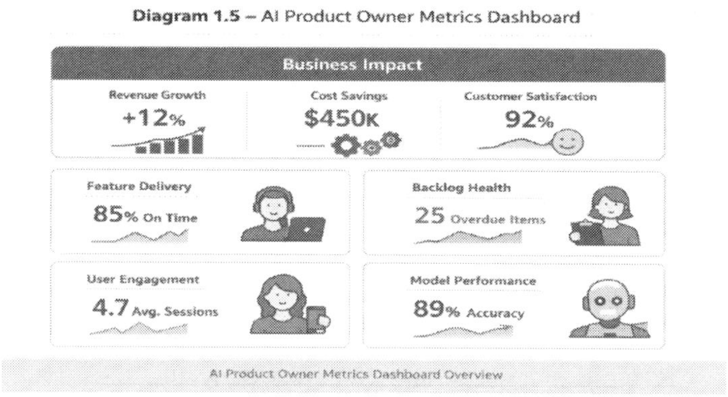

1.10 Manager's Checklist

The following behaviors, artifacts, and decisions characterize an AI Product Owner who is operating at the standard this role requires. Review this checklist at the end of each quarter.

- You can state, without consulting notes, the current production value of each of your model's five key metrics and whether each is within threshold.
- A current model card exists for every model under your ownership, and it was reviewed within the past ninety days.
- A signed AI Product Owner charter exists for every product under your ownership.
- The RACI matrix for your product has been reviewed with all team members and is accessible to all stakeholders.
- Weekly model reviews are on the calendar and have been held consistently for the past three months, with documented decisions for each session.
- Monthly KPI reviews have been held consistently, with attendance from both the technical team and the business owner.
- A postmortem has been completed for every production incident in the past quarter, within the five-business-day commitment.
- The business-to-model outcome mapping has been reviewed in the past ninety days and is still aligned with current business priorities.
- Data governance documentation—data source inventory, privacy impact assessments, data use

agreements—is current and has been reviewed by the compliance contact.

- The escalation ladder has been communicated to all relevant team members, and each level has a confirmed owner with a documented response-time commitment.
- You have communicated model performance and risk status to your senior leader at least once in the past thirty days, in terms that do not require technical expertise to understand.
- You have a documented pilot-to-production pathway for any new model entering the pipeline, with explicit stage-gate criteria.

1.11 Takeaway

The AI Product Owner role is defined by accountability—for data, for models, and for the business outcomes that depend on both. The tools in this chapter—the RACI matrix, the business-to-model outcome mapping, the operating rhythms, the charter, the escalation ladder—are not bureaucratic overhead. They are the mechanisms through which accountability becomes operational clarity. A manager who masters these tools does not just reduce risk; they build the organizational capacity to deliver AI products that are mission-aligned, responsible by design, and capable of sustaining their value across the full product lifecycle.

The chapters that follow build on this foundation. Chapter 2 maps the full AI product lifecycle, from discovery through sunset, including the handoffs, artifacts, and KPIs that keep each stage accountable. Chapter 3 addresses requirements—the translation from fuzzy business goals to testable model specifications that every AI Product Owner must be able to produce—the skills in this book compound. The manager who understands role, accountability, and operating rhythm before engaging with lifecycle, requirements, and data strategy will move faster and make fewer costly mistakes than the manager who encounters each of these challenges for the first time when a production model is already at risk.

2 AI Product Lifecycle: Stages, Handoffs, and KPIs

Marcus had been the engineering lead for the customer-service AI for fourteen months. The model had performed well in staging, and the launch three months ago had felt like a triumph. Now, on a Tuesday morning, his inbox held three escalations from customer success, a terse note from the VP of Operations asking why deflection rates had dropped eleven points in six weeks, and a Slack thread with forty-seven messages in which four different teams were each waiting for one of the other three to take ownership of the fix.

The root cause, when Marcus finally traced it, was straightforward: the production model was receiving a new category of customer queries triggered by a product change that had shipped eight weeks ago. The model had never seen those query patterns. Nobody had flagged the product change to the AI team during launch, and nobody had a monitoring plan that would have automatically caught the drift. There was no defined owner for post-launch model health. The model had slid from high-performer to underperformer in plain sight, and the organization had no instrument panel to notice until customers did.

Marcus's situation is not unusual. It is the predictable outcome of treating AI deployment as a finish line rather than as a transition point in an ongoing operational lifecycle. The chapter that follows provides the manager's framework for preventing exactly this failure: a structured six-stage lifecycle, with explicit artifacts, handoffs, acceptance criteria, and KPIs at every stage.

2.1 Why It Matters

Traditional software products benefit from decades of accumulated lifecycle discipline. Release management, change advisory boards, rollback procedures, and post-mortems are standard operating procedures in mature engineering organizations. AI products arrive inside those same organizations but carry a fundamentally different operational profile: they degrade silently, their failure modes are probabilistic rather than binary, and the boundary between a functioning model and a failing one is a gradient that shifts over weeks and months rather than a hard crash that pages an on-call engineer at 2 a.m.

Without lifecycle discipline, AI products follow a predictable and damaging arc. A prototype earns executive enthusiasm in a demo. Pressure to show results compresses the validation window. The model reaches production without a monitoring plan, a retraining trigger definition, or a clear owner for ongoing health. Within months, performance quietly erodes as real-world data drifts from the distribution the model was trained on. Business stakeholders lose confidence in AI as a delivery vehicle, and the engineering team inherits an unstructured remediation effort with no playbook.

The six-stage lifecycle in this chapter closes that gap. It gives the manager a consistent vocabulary for each phase of an AI product's life, defines the artifacts that must exist before a product advances to the next stage, and ties every stage to the KPIs that make performance visible to both technical and business audiences. When the lifecycle is followed, Marcus's situation does not occur. When it is not, some version of it always does.

Diagram 2.1 – The Six-Stage AI Product Lifecycle

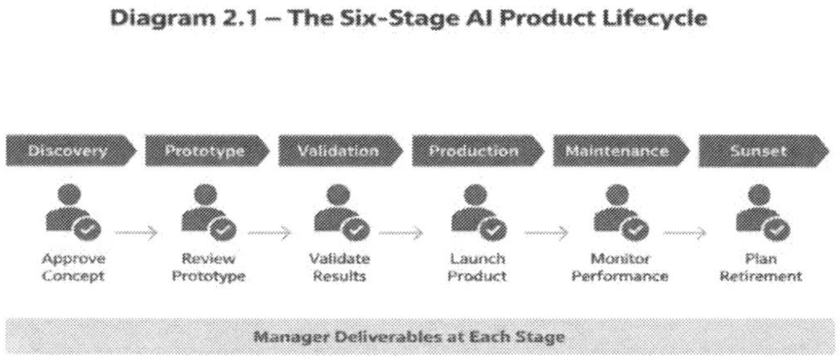

2.2 The Six-Stage AI Product Lifecycle

The lifecycle organizes AI product work into six discrete stages, each with a defined entry condition, a set of work products, and an exit gate. The stages are sequential by default but allow controlled iteration within a stage; no stage should be skipped, and no product should advance past a gate without explicit manager sign-off on the required artifacts.

2.2.1 Stage 1: Discovery

Discovery is the problem-scoping stage. Its purpose is to determine whether an AI approach is warranted at all, and if so, to define the problem precisely enough that data and modeling work can be scoped. The primary output is a problem brief: a one-to-two-page document that states the business goal in measurable terms, describes the data expected to be available, identifies the stakeholders who must be involved, and names the success metric that will govern the project.

Discovery ends when the manager can answer four questions without ambiguity. First, what business outcome is the AI expected to improve, and by how much? Second, what data exist today, and are they sufficient in volume, quality, and labeling coverage? Third, who owns the downstream prediction or output, and what will they do if the model is wrong? Fourth, what does "good enough to deploy" mean in measurable terms? A product that cannot clear these four questions in Discovery has no business entering Prototype.

Discovery typically runs one to three weeks. The manager's role is to facilitate the problem brief, align stakeholders on the success metric before any modeling begins, and prevent the team from falling in love with a technical approach before the problem is fully understood. The most common Discovery failure mode is a vague or moving success metric: "make it better" is not a requirement; "increase first-contact resolution rate from 58 percent to 70 percent within ninety days of deployment" is.

2.2.2 Stage 2: Prototype

Prototype is the rapid experimentation stage. Its purpose is to test whether the problem is technically tractable within the constraints established in Discovery. The team builds a minimal working model—not a production candidate—against a curated slice of training data. The primary output is a prototype evaluation report documenting the model approach, baseline performance against the agreed-upon success metric, known failure modes, and a preliminary estimate of the additional data and engineering effort required to reach production quality.

The Prototype stage carries a specific managerial risk: it is the stage most likely to be cut short by enthusiasm or extended indefinitely by scope creep. A prototype that performs well in a demo can generate internal momentum that bypasses Validation entirely. The manager's job is to enforce the gate. A promising prototype is evidence of feasibility, not evidence of production readiness. It is explicitly not a green light for deployment, regardless of how impressive the demo appears.

The prototype should run no longer than four to six weeks. If the team cannot demonstrate feasibility in that window, the discovery assumptions should be revisited rather than extending the prototype window. A team that needs twelve weeks to demonstrate a proof of concept is usually solving a harder problem than the one scoped in Discovery, and the manager needs to surface that gap explicitly rather than allowing it to be absorbed silently into the schedule.

2.2.3 Stage 3: Validation

Validation is the stage where the prototype becomes a serious candidate for production. The team expands the dataset to representative production-scale volumes, establishes rigorous evaluation protocols including holdout test sets and, where applicable, offline human evaluation, and stress-tests the model against known edge cases and failure scenarios. The primary output is a model card—a structured document that describes the model's intended use, performance across population subgroups, known limitations, and monitoring requirements.

Validation also encompasses the non-technical readiness work that is frequently overlooked. Legal and compliance review should occur during Validation, not after launch. Data privacy impact assessments, if required, belong here. User acceptance testing with a representative sample of end users should happen in Validation, with results documented. The manager should treat a skipped Validation as a significant organizational risk rather than a scheduling convenience.

The exit artifact for Validation is a staging report: a structured summary that covers model performance against acceptance criteria, compliance clearances obtained, outstanding risks with mitigations, and a go/no-go recommendation. The staging report is the primary input to the Stage 3 gate review. No model advances to Production without a completed staging report and an explicit manager decision.

2.2.4 Stage 4: Production

Production is not a state of completion; it is a state of operational accountability. When a model enters production, the organization takes on a standing commitment to monitor its health, respond to incidents, and manage its performance over time. The entry artifact for Production is a monitoring plan: a document that names the business and model metrics that will be tracked, defines alert thresholds, identifies the on-call owner, and specifies the retraining trigger conditions.

Production deployments should follow a staged rollout pattern: shadow mode or canary deployment for the first cohort, with explicit criteria for expanding to full traffic. Shadow mode runs the model on live inputs. Still, it routes outputs to a log rather than presenting them to users, allowing performance validation under real-world traffic conditions without exposing customers. Canary deployment routes a defined percentage of live traffic to the new model, with automatic rollback if KPIs breach defined thresholds.

The first thirty days of Production are the highest-risk period. Real-world data distributions almost always differ from the training distribution in ways that only become visible under live traffic. The manager should maintain a heightened review cadence during this period—daily monitoring check-ins rather than weekly—and should define explicit escalation paths before deployment begins, not after an incident occurs.

2.2.5 **Stage 5: Maintenance**

Maintenance is the operational steady state. The model is in production and performing within acceptable bounds, but ongoing work is required to maintain that performance. Maintenance activities include monitoring KPI trends, investigating drift alerts, executing scheduled retraining cycles, managing data pipeline health, and incorporating feedback from downstream users and human reviewers. The primary management artifact in Maintenance is the monitoring plan, which should be reviewed and updated at least quarterly.

Maintenance is the stage that is most frequently underestimated in planning and most frequently underfunded in budgeting. Organizations that treat AI deployment as a capital project with a defined end date discover in Maintenance that the model requires ongoing engineering capacity that was never allocated. The manager's planning discipline in earlier stages—specifically, forcing explicit acknowledgment of Maintenance costs during Discovery—is the primary mitigation for this failure mode.

Maintenance transitions to a Sunset review when one or more of three conditions apply: the model's performance has degraded beyond the threshold at which retraining can economically restore it; a superior replacement architecture has been validated; or the business need the model addresses has changed materially enough that the original success metric is no longer relevant. The Maintenance-to-Sunset transition should be initiated by the manager, not triggered by an incident.

2.2.6 Stage 6: Sunset

Sunset is the disciplined retirement of a model. It requires the same planning rigor as a new deployment. The primary artifacts are a sunset plan and a knowledge capture document. The sunset plan covers the migration path for any workflows that depend on the model, the communication schedule for affected users and stakeholders, the data archiving or deletion requirements driven by privacy policy, and the timeline for decommissioning the model's infrastructure.

The knowledge capture document is often skipped, but it represents significant organizational value. It records the decisions made during the model's life, the failure modes that were encountered, the monitoring thresholds that proved reliable, and the data characteristics that drove the most meaningful performance changes. This institutional knowledge informs the Discovery stage of the successor model and reduces the cost of future projects in the same domain.

Sunset is not a failure. A model that reached Production, served its intended purpose for a defined period, was retired in an orderly way, and whose lessons were captured is a lifecycle success. The failure mode is not Sunset—it is the model that is never formally retired, continues consuming infrastructure and generating unreliable outputs after its operational context has changed, and for which nobody feels accountable for decommissioning because nobody owns the end-to-end lifecycle.

2.3 Stage Artifacts and Handoffs

Each stage produces a primary artifact that serves as both the evidence of stage completion and the input to the next stage's kickoff. The handoff artifact disciplines the transition between teams: it prevents the common pattern in which work moves from data science to engineering to operations by informal conversation, without documented assumptions, acceptance criteria, or risk disclosures.

Stage	Primary Artifact	Key Contents	Receives / Passes To
Discovery	Problem Brief	Business goal, success metric, data inventory, stakeholder map	Sponsor → Data Science
Prototype	Prototype Evaluation Report	Model approach, baseline metrics, failure modes, effort estimate	Data Science → Manager Gate
Validation	Model Card + Staging Report	Performance, subgroup analysis, compliance clearances, risks	Engineering → Legal → Manager Gate
Production	Monitoring Plan	KPI targets, alert thresholds, on-call owner, retraining triggers	Manager → Operations
Maintenance	Updated Monitoring Plan	Quarterly drift review, threshold adjustments, and incident log	Operations → Manager Review
Sunset	Sunset Plan + Knowledge Capture	Migration path, decommission timeline, lessons learned	Manager → Successor Team

The handoff artifact is not a bureaucratic hurdle; it is a communication instrument. A well-written problem brief prevents the data science team from spending four weeks optimizing the wrong metric. A well-written model card ensures that the operations team knows which failure modes to monitor before a live incident exposes them. The manager who demands these artifacts at each gate is not adding friction to the process—they are removing the ambiguity that makes later stages expensive.

Handoffs fail for two common reasons. The first is incompleteness: the artifact exists but is too thin to serve its purpose. A two-sentence model card is not a model card; it is a cover letter for an undocumented model. The manager must set and enforce quality standards for each artifact type, and those standards should be defined in the team's operating norms, not improvised at review time. The second failure mode is timing: the artifact is produced after the handoff. The problem brief written after the prototype has been built is not a planning document; it is a post-hoc rationalization. The disciplines only work when applied in sequence.

2.4 Manager's Deliverables and Acceptance Criteria per Stage

The manager plays a distinct role at each stage, going beyond approving the primary artifact. The following section defines the manager's specific deliverable—the decision or action that only the manager can complete—and the acceptance criteria that must be satisfied before advancing to the next stage. These are not aspirational standards; they are the minimum conditions for responsible progression.

2.4.1 Discovery: Align on the Success Metric

The manager's deliverable in Discovery is an aligned, written success metric accepted by the business sponsor, the data science lead, and the product owner. "Accepted" means formally acknowledged in writing, not verbally agreed in a meeting that is later recalled differently by each participant. The acceptance criterion for advancing from Discovery is that the problem brief exists, the success metric is specific and measurable, and at least one person outside the AI team has confirmed that achieving the metric would constitute a meaningful business outcome.

2.4.2 Prototype: Gate the Feasibility Decision

The manager's deliverable at the Prototype gate is a documented feasibility decision: proceed to Validation, return to Discovery with revised assumptions, or terminate the project. This decision must be documented even when it is a procedure—the rationale, the performance against the success metric, and the known gaps that Validation will need to address. The acceptance criterion is a prototype evaluation report that demonstrates measurable performance on the agreed metric, not merely a demonstration on cherry-picked examples.

2.4.3 Validation: Clear the Compliance and Quality Gates

The manager's deliverable in Validation is evidence of the completion of the compliance review and includes a staging report with a go/no-go recommendation. The acceptance criterion for Production entry is that the model has been evaluated on a holdout dataset that the manager can confirm was not used in training, that legal and compliance review is documented, and that the monitoring plan has been drafted—not just planned. A model that passes technical evaluation but lacks a monitoring plan has not cleared the Validation gate.

2.4.4 Production: Confirm the Operational Handoff

The manager's deliverable at Production entry is a confirmed operational handoff: a named on-call owner, a documented alert runbook, and an explicit first-thirty-day review schedule. The acceptance criterion is that the monitoring plan has been reviewed and acknowledged by the operations team, that alert channels have been tested, and that the escalation path from alert to incident response is unambiguous. The manager should be able to answer, without consulting documentation, "If this model fails at 11 p.m. on a Saturday, who gets paged and what do they do?"

2.4.5 Maintenance: Sustain the Review Cadence

The manager's deliverable in Maintenance is a sustained review cadence. This means a standing calendar item for model health review, a quarterly refresh of the monitoring plan, and an annual lifecycle review to assess whether the model should continue in Maintenance or enter a Sunset planning process. The acceptance criterion is that the review cadence is documented in the team's operating norms and has been honored for at least two consecutive cycles since the model's Production entry.

2.4.6 **Sunset: Validate the Migration and Knowledge Capture**

The manager's deliverables at Sunset are a completed migration confirmation and a knowledge capture document. The acceptance criterion is that all downstream workflows have been migrated or formally accepted as discontinued, that the knowledge capture document has been stored in a location accessible to the successor team, and that the model's infrastructure has been decommissioned with evidence. A model that is "sort of retired"—no longer actively maintained but still receiving some traffic—has not cleared the Sunset gate.

Diagram 2.2 – Stage-Gate Decision Points and Manager Deliverables

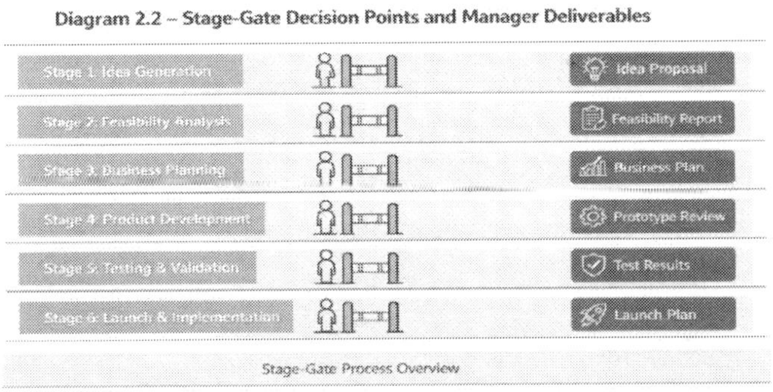

2.5 The Stage-Gate Checklist

The stage-gate checklist is the manager's primary decision instrument at each lifecycle transition. It is designed to be completed in under fifteen minutes for a well-prepared project, and it should feel frictionless when the team has been executing well. When the checklist feels difficult to complete, that difficulty is diagnostic: it indicates gaps in the stage work that the manager needs to address before advancing. The checklist is organized by stage and covers the minimum conditions for a responsible go decision.

2.5.1 Gate 1: Discovery to Prototype

- **Problem Brief complete:** business goal, success metric, and data inventory are documented and stakeholder-reviewed
- **Success metric agreed:** metric is specific, measurable, and accepted in writing by the business sponsor
- **Data availability confirmed:** at least a preliminary data readiness assessment has been completed
- **Stakeholder map complete:** roles, decision rights, and review responsibilities are documented
- **Termination conditions defined:** criteria for stopping the project before reaching Production are documented

2.5.2 Gate 2: Prototype to Validation

- **Prototype evaluation report exists:** model approach, baseline metrics, and known failure modes are documented

- **Feasibility demonstrated:** model achieves measurable performance on the agreed metric, not only on curated examples
- **Data gaps identified:** the report names any data quality or volume gaps that Validation must address
- **Effort estimate provided:** the report includes a rough estimate of what Validation and Production will require
- **Manager's decision documented:** the feasibility decision and its rationale are recorded in writing

2.5.3 Gate 3: Validation to Production

- **Model card complete:** intended use, performance metrics, subgroup analysis, and known limitations documented
- **Staging report complete:** holdout evaluation results, compliance clearances, and go/no-go recommendation documented
- **Legal and compliance review completed:** privacy impact assessment and any required approvals are documented
- **Monitoring plan drafted:** KPI targets, alert thresholds, on-call owner, and retraining triggers defined
- **Staged rollout plan defined:** shadow or canary deployment parameters and success criteria documented

2.5.4 Gate 4: Production Entry

- **Monitoring plan acknowledged:** the operations team has reviewed and confirmed the monitoring plan
- **Alert channels tested:** alert routing has been verified in the target environment
- **Escalation path documented:** named on-call owner and incident response runbook exist
- **First-thirty-day review scheduled:** calendar items exist for daily check-ins during the initial production period
- **Rollback plan confirmed:** the procedure for reverting to the prior state is documented and tested

2.5.5 Gate 5: Maintenance to Sunset Review

- **Performance trend reviewed:** trailing-twelve-month performance data has been reviewed against original acceptance criteria
- **Maintenance cost assessed:** ongoing engineering and infrastructure costs have been compared against the business value delivered
- **Successor options identified:** alternatives to continued Maintenance (rebuild, replacement, retirement) have been evaluated
- **Lifecycle review completed:** annual lifecycle review has been held and documented
- **Sunset trigger condition met:** at least one of the defined sunset trigger conditions has been confirmed

Diagram 2.3 – Stage-Gate Checklist: Go/No-Go Visual Summary

Diagram 2.3 – Stage-Gate Checklist: Go/No-Go Visual Summary

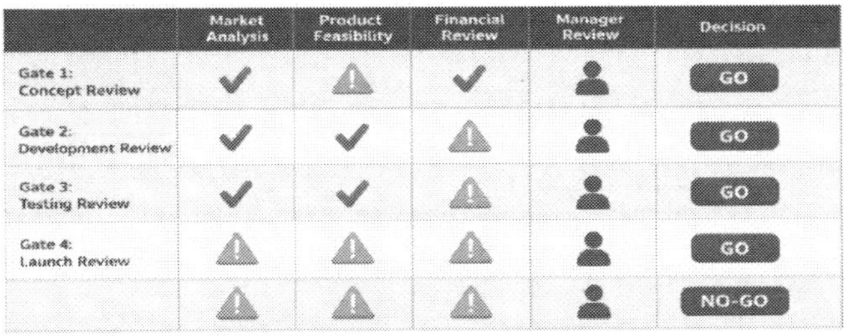

2.6 The Go/No-Go Decision Matrix

The stage-gate checklist establishes minimum conditions; the go/no-go decision matrix provides the framework for evaluating checklist gaps against risk tolerance. Not every gap is equal. A missing stakeholder signature on the problem brief is a process gap that can often be remedied within 24 hours. A missing compliance review at Gate 3 is a risk exposure that should stop the gate, regardless of business pressure to launch. The matrix below structures the manager's risk weighting across four dimensions.

Criterion	Low Risk	Moderate Risk	High Risk	Blocker
Success Metric	Specific, agreed in writing	Defined but not yet agreed	Loosely defined	Absent or contested

Criterion	Low Risk	Moderate Risk	High Risk	Blocker
Data Readiness	Representative, labeled, sufficient volume	Sufficient volume, quality gaps noted	Volume gaps, no quality assessment	Data not available or not assessed
Compliance Review	Completed and documented	In progress, low-sensitivity data	Not started, medium-sensitivity data	Not started, regulated/sensitive data
Monitoring Plan	Complete and ops-acknowledged	Drafted, not yet reviewed	KPIs defined, thresholds missing	Absent
Rollback Plan	Documented and tested	Documented, not yet tested	Partially documented	Absent
Staged Rollout	Shadow or canary plan confirmed	Canary plan drafted	Direct-to-full traffic planned	Not planned for a regulated use case

A gate with a single Blocker criterion should not proceed. A gate with two or more High Risk criteria should be deferred until the risks are resolved or formally accepted in writing by the business sponsor—not accepted verbally, not accepted by the product manager on behalf of the sponsor, but accepted in writing by the person accountable for the business outcome. The purpose of written acceptance is not to assign blame when things go wrong; it is to make the risk visible before deployment so that the accepting party is making an informed decision.

The matrix is a decision aid, not a decision algorithm. A manager who uses it mechanically—accepting every Moderate Risk without considering its specific context—is not exercising the judgment the matrix is designed to support. A moderate-risk data quality gap in an internal routing tool differs from a moderate-risk data quality gap in a model that makes lending recommendations. The same matrix cell can warrant a different response depending on the deployment context's stakes. The manager's obligation is to apply judgment informed by the matrix, not to be replaced by it.

2.7 Lifecycle KPIs That Connect Business and Model Health

KPIs serve two distinct audiences in an AI product lifecycle: the business stakeholders who care about outcomes, and the technical team that needs to detect model degradation before it becomes a business problem. Most organizations track one or the other, but rarely both in a unified framework. The manager's role is to hold both sets of metrics simultaneously, because the failure mode of tracking only business KPIs is Marcus's situation, and the failure mode of tracking only model metrics is a technically healthy model that nobody can connect to a business outcome.

The following framework organizes lifecycle KPIs by stage and by audience. Business KPIs appear at every stage but are measured differently: in Discovery, they are targets; in Prototype, they are early estimates; in Validation, they are acceptance criteria; in Production and Maintenance, they are the primary health signals. Model metrics appear from Prototype onward and become the primary degradation early-warning system in Maintenance.

2.7.1 Discovery and Prototype KPIs: Establishing Targets

In Discovery, the KPI work is definitional. The manager and business sponsor must agree on a primary business KPI—the number that, if achieved, constitutes success—before any modeling begins. Common primary KPIs include conversion rate improvement, cost-per-transaction reduction, first-contact resolution rate, cycle time reduction, and error rate reduction. The target value should be based on business analysis: what outcome change would justify the investment? A target pulled from a vendor benchmark or an internal wish list without reference to the cost of achieving it is not a KPI; it is an aspiration.

Secondary KPIs should also be defined in Discovery. These are the guardrail metrics: the measures that must not degrade while the primary KPI improves. If a customer service AI improves first-contact resolution but increases average handle time by forty percent, that is not a net improvement. The guardrail metric—average handle time within a defined bound—is what prevents optimization on the primary KPI from creating damage elsewhere. Every Discovery document should name at least one guardrail KPI alongside the primary success metric.

2.7.2 Validation KPIs: Acceptance Criteria

In Validation, KPIs become acceptance criteria. The model must meet the primary business KPI target on the holdout evaluation dataset and must not exceed any guardrail KPI thresholds. In addition, Validation introduces model-specific metrics that do not appear in the business KPI framework: precision and recall (or their equivalents for the task type), confidence calibration, latency at the 95th and 99th percentiles, and inference cost per prediction.

Confidence calibration is worth particular attention because it is frequently omitted from Validation evaluation. A model that reports 90 percent confidence in its predictions should be correct approximately 90 percent of the time on that subset of predictions. A model that is systematically overconfident—reporting high confidence on predictions that are frequently wrong—is especially dangerous in workflows where human reviewers use the model's confidence score to decide whether to review a prediction. Validation must include a calibration analysis, and the model card must document the calibration results.

Diagram 2.4 – KPI Framework: Business and Model Metrics by Lifecycle Stage

Diagram 2.4 – KPI Framework: Business and Model Metrics by Lifecycle Stage

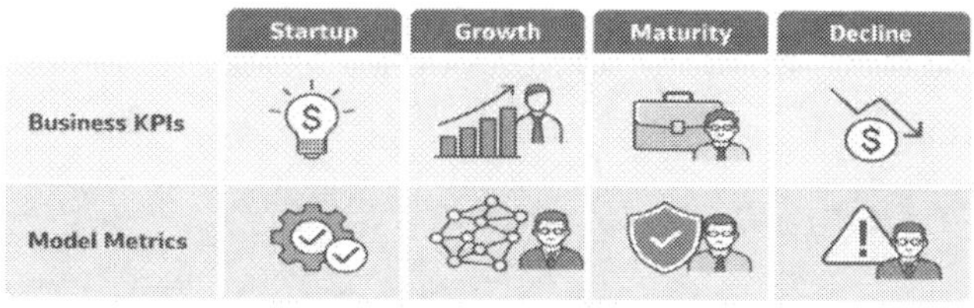

Key performance indicators and model metrics tracked through the lifecycle stages of a business.

2.7.3 Production and Maintenance KPIs: The Four Essential Pairs

In Production and Maintenance, the manager needs four paired KPI sets. Each pair links a business-level signal to a model-health indicator so that the manager can trace a shift in business performance to its model-level cause, or detect a model-level problem before it produces a business-level impact. The four pairs are as follows.

The first pair is conversion rate and precision. Conversion rate—whether the customer took the desired action following an AI recommendation—is the business signal. Precision—the proportion of correct positive predictions—is the model signal. When conversion rate drops but precision holds, the cause is likely upstream of the model (user experience, offer quality, traffic mix). When precision drops, it typically precedes a drop in conversion rates by days or weeks. Monitoring both allows the manager to distinguish model problems from business problems.

The second pair is cost savings and drift rate. Cost savings—whether the AI is reducing operational cost at the projected rate—is the business signal. Drift rate—the statistical divergence between the current input distribution and the training distribution—is the model signal. Drift does not always cause immediate degradation, but it always precedes it. A model with a stable cost-savings trajectory but an elevated, rising drift rate will fail within weeks to months. The manager who monitors drift rate can intervene proactively; the manager who monitors only cost savings cannot.

The third pair is error budget and recall. Error budget—the defined acceptable rate of incorrect or missed outputs expressed in business terms (failed transactions per thousand, mis-routed tickets per day)—is the business signal. Recall—the proportion of actual positive cases that the model identified—is the model signal. Recall degradation is the most common early signal of a model that is becoming overly conservative as the data distribution shifts, and it often appears in the error budget as an increase in missed cases before it registers as a business-level complaint.

The fourth pair consists of the customer satisfaction score and the 95th-percentile latency. Customer satisfaction is the business signal; latency is the model signal. The relationship is not always linear, but sustained latency above a threshold—typically 500 milliseconds for synchronous customer-facing applications, though the specific threshold must be calibrated to the application—is reliably predictive of a decline in satisfaction. Infrastructure changes, model size increases during retraining, and increases in concurrent request volume can all cause latency drift that is invisible in average latency measurements but visible at the 95th percentile.

2.8 Combining Business Impact with Model Metrics

The four KPI pairs described above are most useful when reviewed together, not in separate dashboards owned by different teams. The most common failure mode in AI monitoring is a bifurcated measurement environment: the data science team monitors model metrics in one system, the business team monitors business KPIs in another, and neither team has a view that connects the two. When a business KPI degrades, the business team escalates to the data science team, who then begins a manual investigation to determine whether the cause is a model problem or a business problem. This investigation takes time, further degrading the business outcome.

The unified monitoring dashboard—a single view that presents both business KPIs and model metrics for the same time period, with the four paired relationships visible—is the operational instrument that prevents this failure. The manager does not need to be the person who builds this dashboard, but must be the one who demands it and refuses to accept a Production deployment without it. A team that argues that combined monitoring is too complex to implement has not been held to the standard. Combined monitoring is an engineering task of moderate complexity; it is not an unsolved problem.

The unified dashboard should also surface the causal direction of observed changes. A drop in conversion rate accompanied by stable model precision and a recent product change in the upstream user experience points to a product problem, not a model problem. A drop in conversion rate accompanied by declining precision and elevated input drift points to a model problem. The dashboard should make these distinctions traceable without requiring a manual investigation for each incident. Annotations for product releases, infrastructure changes, and external events (seasonal patterns, regulatory changes, marketing campaigns) should be overlaid on the metric time series to make correlations immediately visible.

Diagram 2.5 – Unified Monitoring Dashboard: Business KPIs and Model Metrics

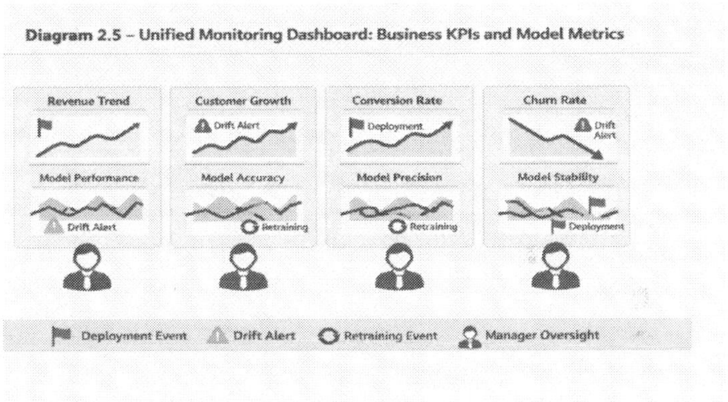

2.9 Vignette: A Disciplined Stage-Gate Prevents a Premature Production Launch

Priya was the product manager for a claims triage AI at a mid-size insurance carrier. The model was designed to classify incoming claims by complexity, routing straightforward cases to automated processing and complex or ambiguous cases to senior adjusters. The prototype had performed well in internal testing, and the carrier's claims VP had publicly committed to a Q3 launch in a presentation to the board. By early July, the engineering team was ready to push to production. The prototype evaluation looked strong. The VP was asking for a launch date.

Priya ran the Gate 3 checklist before confirming the date. The model card was complete. The staging report showed strong precision on the holdout set. But two items on the checklist were incomplete: the compliance review had not addressed the carrier's state regulatory obligations regarding automated claims decision-making in two of the seven states where the model would operate, and the monitoring plan included KPI targets but no alert thresholds or a named on-call owner for model health. On the go/no-go matrix, the compliance gap was a Blocker for the affected states, and the monitoring plan gap was High Risk.

Priya brought the gap analysis to the VP in a brief written summary. She did not frame it as a delay; she framed it as a scope reduction. The model could launch on schedule in the five states with completed compliance clearances, with the remaining two states following six weeks later, after regulatory review was complete. The monitoring plan would be completed before any production traffic was routed, with the on-call owner confirmed and the alert runbook tested. The VP agreed. The phased launch proceeded on the original date in five states, with no incidents. The remaining two states launched seven weeks later, within the projected window.

The counterfactual is straightforward: a launch in all seven states without completed regulatory review would have exposed the carrier to compliance risk that the compliance team estimated at material financial and reputational levels. The monitoring plan gap would have produced exactly the situation Marcus encountered. Priya's discipline at Gate 3 was not a performance of caution; it was the routine application of a lifecycle framework that her team had internalized. The checklist did not slow the project—it prevented a problem that would have cost far more time to remediate than the six-week delay.

2.10 Manager's Checklist

The following checklist consolidates the manager's key lifecycle responsibilities into a single reference. It is designed to be reviewed at the start of a new AI product initiative and revisited at each stage gate.

- Confirm the problem brief exists and the success metric is agreed in writing before Prototype begins
- Enforce the four-to-six-week Prototype window; do not allow indefinite extension
- Require a prototype evaluation report with documented failure modes before advancing to Validation
- Ensure compliance review is initiated at the start of Validation, not after staging is complete
- Demand a completed model card and staging report with a go/no-go recommendation before Gate 3

- Confirm a monitoring plan with named alert thresholds, on-call owner, and retraining triggers before Production entry
- Schedule a staged rollout (shadow or canary) as the default Production entry path
- Maintain daily model health check-ins for the first thirty days of Production
- Establish a standing quarterly monitoring plan review and annual lifecycle review
- Initiate Sunset planning when maintenance cost exceeds value delivered or a trigger condition is met
- Complete a knowledge capture document before decommissioning any model
- Track all four KPI pairs (conversion/precision, cost savings/drift, error budget/recall, satisfaction/latency) in a unified dashboard
- Require written acceptance from the business sponsor for any High Risk gate gap before advancing
- Document all go/no-go decisions and their rationale in a durable record accessible to the successor team

2.11 Takeaway

The AI product lifecycle is not a process tax. It is the operational structure that converts a promising model into a reliable product and keeps it reliable over time. Each of the six stages—Discovery, Prototype, Validation, Production, Maintenance, and Sunset—exists because the failure modes it prevents are real, recurring, and expensive. The artifacts and handoffs are not documentation for its own sake; they are the communication instruments that prevent misaligned assumptions from compounding across team boundaries.

The stage-gate checklist and go/no-go matrix are not bureaucratic checkpoints; they are the decisional instruments that make the manager's role visible and actionable at every transition.

Marcus's problem—a model in production without a monitoring plan or an accountable owner—is a lifecycle problem. It does not require a technical solution; it requires an organizational discipline that the manager is uniquely positioned to enforce. The manager who runs the Gate 3 checklist and refuses to advance a model without a monitoring plan is not adding friction. They are performing the function that lifecycle discipline is designed to support: preventing the predictable failures that emerge when capable teams move fast without the structural safeguards that turn speed into sustainable delivery.

The KPI framework reinforces this discipline at the operational level. Business KPIs connect the model's performance to the outcomes that justify its existence. Model metrics provide an early warning system that enables proactive management. Tracking both in a unified dashboard and holding both to explicit thresholds are what separate organizations that manage AI products from those that do not. The lifecycle, the gates, and the KPIs are the tools. The manager's judgment—applied consistently, documented transparently, and revisited at each transition—is what makes them work.

3 Requirements: From Outcomes to Measurable Specs

The message arrived on a Friday afternoon: the customer-service chatbot was going live in two weeks, the steering committee was pleased with the demos, and the engineering team had hit every sprint milestone. Six weeks later, a second message arrived — this one considerably less upbeat. Users complained that the chatbot was confidently providing incorrect answers. Return policy details were invented. Escalation procedures were garbled. One regional manager counted fourteen factual errors in a single shift. When the incident review convened, a painful truth emerged: nobody had ever defined what a "good enough" response looked like. The business had asked for a chatbot that "improves customer satisfaction." Engineering had interpreted that as a reduction in average handle time. The data team had optimized for response coherence. Nobody had set a hallucination threshold. Nobody had agreed on what failure looked like. The product had shipped on time and immediately started failing.

This scenario plays out across industries with remarkable consistency. A goal that sounds concrete in a boardroom — "improve customer satisfaction," "reduce manual review," "accelerate onboarding" — dissolves into ambiguity the moment it touches a model training run. Each team fills the gap with its own interpretation, and those interpretations silently diverge. By the time the system reaches real users, the gap has widened into a chasm that no sprint review caught, because no sprint review was looking at the right thing.

Chapter 3 is about closing that gap before it opens. It provides a structured approach to translating business goals into model requirements that are specific enough to test, precise enough to negotiate, and durable enough to survive the hand-off from discovery to production. The tools here are not exotic. They are discipline, vocabulary, and a willingness to make the conversation about numbers rather than adjectives.

3.1 Why It Matters

Fuzzy requirements are the single most reliable predictor of AI project rework. In conventional software, a vague requirement produces a feature that looks wrong — visible to anyone who clicks through the UI. In AI systems, a vague requirement yields a model that performs at an unknown level against an unstated standard. The failure is invisible until it touches real data, real users, or real consequences. By then, weeks of labeling work, fine-tuning cycles, and infrastructure buildout have been spent on a target that was never properly defined.

The manager's role in requirements is not to write the metric formulas — that's the data scientist's job. The manager's role is to force the conversation, hold the parties accountable to testable language, and refuse to move the project forward until the definitions are in writing. This is not bureaucratic obstruction. It is the single most cost-effective intervention available to a product owner. A two-hour requirements workshop that surfaces disagreement about acceptable precision rates will save far more time than the sprint review that discovers the same disagreement six weeks later.

There is also a stakeholder dimension that goes beyond engineering efficiency. Business partners who sign off on vague goals will later dispute results measured against those goals. Legal and compliance teams that never saw a hallucination-tolerance threshold will assert, after an incident, that they would never have approved such a threshold. Precise requirements create a shared record of what was agreed, when, and by whom. They transform "the model failed" from a blame conversation into a diagnostic conversation.

3.2 Translating Business Goals to Model Requirements

The gap between "improve customer satisfaction" and a measurable model target is wider than most managers expect, and navigating it requires a deliberate translation process rather than a single intuitive leap. The translation proceeds through at least three layers: the business outcome, the operational metric that drives it, and the model behavior that produces that metric.

Consider the customer satisfaction goal. The operational metric that most directly influences it might be first-contact resolution — the percentage of customer inquiries resolved without escalation or callback. The model behavior that drives first-contact resolution might be response accuracy, defined as the percentage of responses that correctly state policy, pricing, or procedure. That gives you something to measure: response accuracy of 95% or higher on a held-out test set drawn from the last ninety days of support tickets. That is a requirement. "Improve customer satisfaction" is a wish.

3.2.1 The Three-Layer Translation Model

The three-layer structure — business outcome, operational metric, model behavior — is a forcing function, not a formula. In some cases, the operational metric is already well-defined, and the work is finding the model behavior that drives it. In others, the business outcome is contested, and the manager must negotiate which operational metric the organization will use as a proxy for it. Either way, the translation process surfaces assumptions that would otherwise remain hidden.

A useful exercise is to ask the business stakeholder: "If the model achieves X on metric Y, are you prepared to declare the project successful?" If the answer is yes, you have a requirement. If the answer is "it depends," press on what it depends on. The dependencies are the real requirements. Common dependencies include performance by user segment (the model must perform equally well for enterprise customers and small-business customers), performance by topic category (high accuracy on billing questions is more important than on general FAQ), and performance under load (the model must maintain latency targets at peak traffic).

Diagram 3.1 – Business Goal to Model Requirement Translation Layers

Diagram 3.1 – Business Goal to Model Requirement Translation Layers

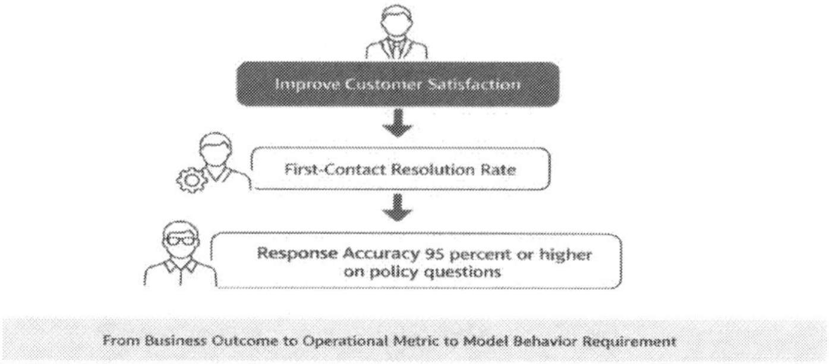

Document the translation path explicitly. In your PRD addendum (detailed later in this chapter), include not just the target metric but the reasoning chain: what business outcome the metric serves, what operational variable it proxies, and what assumptions would invalidate the proxy. This creates an audit trail and gives future reviewers context for understanding why a particular threshold was chosen.

3.2.2 Common Translation Traps

Several patterns reliably produce poor requirements. The first is optimizing for what is easy to measure rather than what matters. Latency is easy to measure; factual accuracy requires human judgment and is much harder to assess. Teams under pressure to show a metric will gravitate toward latency, throughput, or format compliance — and call the job done. Guard against this by anchoring the requirements conversation to the operational metric that the business actually cares about, even if measuring it is labor-intensive.

The second trap is setting requirements against the wrong population. A model that achieves 95% accuracy on a curated evaluation set built from clean, well-formatted support tickets may perform far worse on the messy, misspelled, code-switched messages that arrive during peak hours. Requirements must specify not just the target but the population on which the target applies.

The third trap is confusing capability with requirement. A model may be capable of handling twelve languages. That capability is irrelevant unless the requirement specifies which languages must be supported, at what accuracy level, and what happens when the model encounters a language outside that set. An unspecified capability is a liability because it creates user expectations that the model may not consistently meet.

3.3 Defining Target Metrics

Target metrics translate model behavior requirements into numbers. The manager does not need to derive these numbers from first principles, but does need to understand what they measure, what they hide, and how to interpret them in a business context. The following metrics form the core vocabulary for AI requirements conversations.

3.3.1 Classification Metrics: Precision, Recall, and F1

Precision measures the fraction of correct positive predictions. A fraud-detection model with 90% precision makes ten false accusations for every ninety correct ones. Recall measures the fraction of actual positives that the model catches. A model with 90% recall misses ten fraudulent transactions for every ninety it flags. F1 is the harmonic mean of precision and recall, and it is useful when the cost of missing a positive is roughly equal to the cost of a false positive.

The business relevance of the precision-recall tradeoff is that the two metrics pull in opposite directions, and the right balance is a business decision, not a technical one. In fraud detection, false negatives (missed fraud) are usually more costly than false positives (legitimate transactions flagged for review), so requirements should favor recall. In content moderation for a professional platform, false positives (legitimate content incorrectly removed) may damage user trust more severely than false negatives, so precision takes priority. The manager's job is to make this tradeoff explicit before the model is trained, not after.

3.3.2 Latency and Throughput

Latency requirements specify the maximum acceptable response time for a given request percentile. A typical production requirement might read: "p95 latency under 800 milliseconds for synchronous inference requests during business hours, and p99 latency under 2,000 milliseconds." The percentile framing matters because average latency can appear acceptable while a significant tail of slow requests degrades the user experience.

Throughput requirements specify the volume the system must handle, such as requests per second, requests per day, or concurrent sessions. Throughput interacts with cost: a model that handles 100 requests per second at acceptable latency may require significantly more infrastructure than one handling 10 requests per second, and the cost difference must be anticipated in the requirements, not discovered during capacity planning.

3.3.3 Cost Per Inference

Cost per inference is frequently omitted from early requirements conversations, only to become a crisis at scale. A model that costs $0.08 per inference looks cheap in a pilot of 500 requests per month and becomes a $480,000 annual line item at 500,000 requests per month. Requirements should specify the maximum acceptable cost per inference at the expected production volume, and should include a trigger for re-evaluation if volume projections change significantly.

For LLM-based systems, cost per inference is driven primarily by token count: the number of tokens in the prompt (including any system instructions and retrieved context) plus the number of tokens in the response. Requirements for LLM systems should therefore include maximum prompt and response lengths, and the expected distribution of request complexity.

Diagram 3.2 – Precision-Recall Tradeoff Decision Matrix

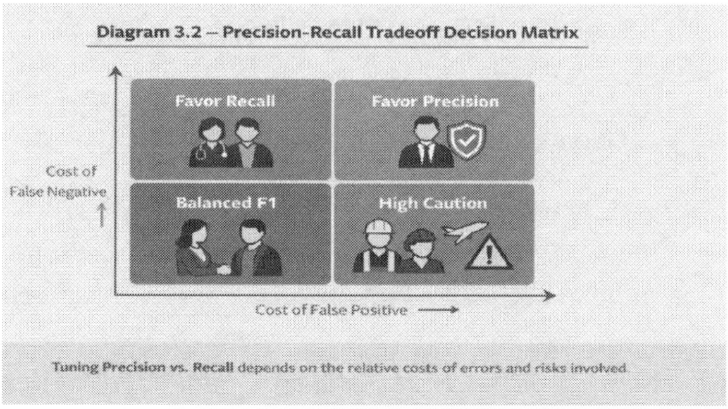

3.3.4 Accuracy and Error Rate Baselines

Target metrics are only meaningful relative to a baseline. Before setting a precision target of 92%, establish what the current baseline is: the accuracy of the manual process being replaced, the performance of a simple rules-based system, or the performance of a competitor's model if benchmarks are available. A model that achieves 92% precision on a problem where the baseline is 88% delivers four percentage points of improvement — meaningful but modest. A model achieving 92% on a problem where the manual baseline is 70% is delivering transformative accuracy gains.

Baselines also provide the floor below which a model should not be deployed. If the manual process achieves 80% accuracy, a model that achieves 75% accuracy is not a product — it is a regression. This should be stated in the requirements as a go/no-go threshold, not left to informal judgment at launch.

3.4 Acceptable Failure Modes and Edge Cases

Defining what a model is allowed to get wrong is as important as defining what it must get right. This is an uncomfortable conversation because it requires the business to endorse imperfection explicitly, and organizations are culturally conditioned to avoid doing so. Push through the discomfort. A model that has no defined failure tolerance will be held to an implicit standard of perfection, and that standard will be invoked selectively whenever a high-profile error occurs.

Acceptable failure modes should be documented in two dimensions: type and rate. The type dimension asks what category of error is acceptable. A customer-service chatbot might be permitted to fail gracefully on requests outside its documented scope (e.g., misrouting to a human agent), but not to provide factually incorrect information about return policies. The rate dimension asks how often each failure type may occur. Graceful failures on out-of-scope requests might be acceptable at any rate; factual errors on core policy topics might be acceptable at no more than one per thousand interactions.

3.4.1 Error Taxonomy

Develop a simple error taxonomy before training begins. The taxonomy should be specific enough to distinguish consequential errors from benign ones, but simple enough to be maintained by a non-specialist. A practical structure for most AI applications has three levels:

critical errors (errors with direct business, legal, or safety consequences), significant errors (errors that degrade user experience or require manual remediation), and minor errors (errors that are noticeable but self-correcting or trivially fixable).

The critical error category should carry zero-tolerance treatment or near-zero tolerance. Significant errors should be subject to a rate threshold (e.g., fewer than 2% of responses) and a remediation SLA (e.g., critical errors escalated within 4 hours, root cause analysis within 48 hours). Minor errors should be tracked but not treated as blockers.

3.4.2 Edge Case Requirements

Edge cases are input conditions that fall outside the model's primary training distribution but that real users will nevertheless generate. They require explicit requirements because models degrade in predictable ways at distribution boundaries, and that degradation may be acceptable or unacceptable depending on the context.

Common edge-case categories include inputs in languages not in the primary training set; inputs with unusually long or short text; inputs containing special characters, code, or structured data; inputs that reference events after the model's training cutoff; and adversarial inputs designed to elicit specific behaviors. Requirements should specify, for each edge case category, whether the model must handle it (with what accuracy), must detect it and route it elsewhere, or may reject it with an explicit error message.

Diagram 3.3 – Error Taxonomy and Tolerance Thresholds

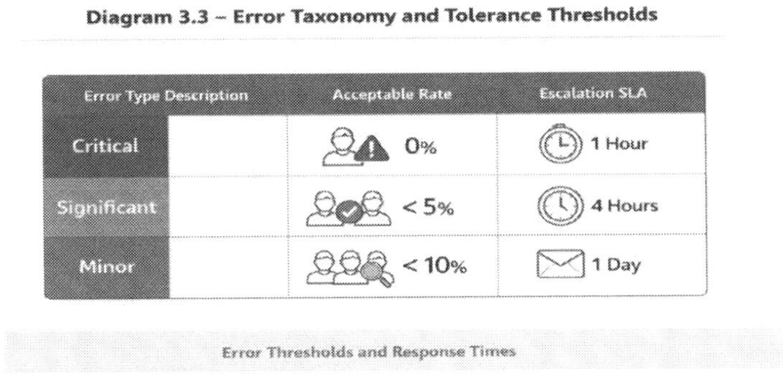

Diagram 3.3 – Error Taxonomy and Tolerance Thresholds

Error Type Description		Acceptable Rate	Escalation SLA
Critical		0%	1 Hour
Significant		< 5%	4 Hours
Minor		< 10%	1 Day

Error Thresholds and Response Times

3.5 Requirements for LLMs and RAG Systems

Large language models and retrieval-augmented generation systems introduce requirements categories that have no direct equivalent in traditional supervised learning. Managers working on LLM-based products need a working vocabulary for these categories to conduct effective requirements conversations with engineering and data teams.

3.5.1 Response Accuracy and Hallucination Tolerance

Hallucination — the generation of confident, plausible-sounding statements that are factually incorrect — is the defining failure mode of LLMs. Unlike a classification error, which produces a discrete wrong answer, a hallucination produces wrong content that looks right. This makes it difficult to catch without human review or a specialized evaluation pipeline, and the business consequences can be severe in any context where users act on model output.

Requirements for hallucination tolerance should specify the maximum acceptable rate of factually incorrect statements per unit of output, the evaluation method (human review, automated fact-checking against a reference corpus, or consistency testing), and the sample size and cadence of evaluation. For high-stakes applications — legal, medical, financial, regulatory — hallucination tolerance should be near zero, and the evaluation burden is therefore high. For lower-stakes applications, a higher tolerance may be acceptable, but it must still be written down.

RAG systems introduce a distinct accuracy requirement: retrieval accuracy, which measures whether the system surfaces the correct source documents before generating a response. A RAG system with 98% generation accuracy but 70% retrieval accuracy will fail to find relevant documents 30% of the time, producing either hallucinated answers or unhelpful refusals. Requirements for RAG systems should include retrieval recall (what percentage of relevant documents are retrieved), retrieval precision (what percentage of retrieved documents are relevant), and response groundedness (what percentage of claims in the response are supported by retrieved documents).

3.5.2 Latency in LLM Systems

LLM latency has two components that must be specified separately: time to first token (the delay before the response begins streaming) and time to complete response (the total time for the full response to appear). Users tolerate streaming responses longer than they tolerate blank screens, so a system that streams at 50 tokens per second with a 300-millisecond first-token latency may feel faster than one that delivers the complete response at 800 milliseconds without streaming. Requirements should specify both the components and the expected response-length distribution.

For synchronous, non-streaming use cases (such as automated document processing), only end-to-end latency matters, and it must be specified at the appropriate percentile given the workload volume and variability.

3.5.3 Privacy and Data Handling Constraints

LLM systems operating on enterprise data introduce privacy requirements that must be specified before architecture decisions are made. These requirements fall into three categories: data at rest (what data may be stored in the LLM provider's infrastructure, and for how long), data in transit (whether requests must be encrypted, and under what protocol), and data retention for fine-tuning (whether request-response pairs may be used to improve the model, and under what consent framework).

Many cloud LLM providers offer zero-retention API options that process requests without logging them. These options carry a higher cost and sometimes lower availability. The requirements document should specify whether zero retention is required, whether the application involves regulated data categories (PII, PHI, financial data, attorney-client communications), and what the data residency constraints are (which geographic regions the data may traverse or be stored in).

Diagram 3.4 – RAG System Requirements Framework

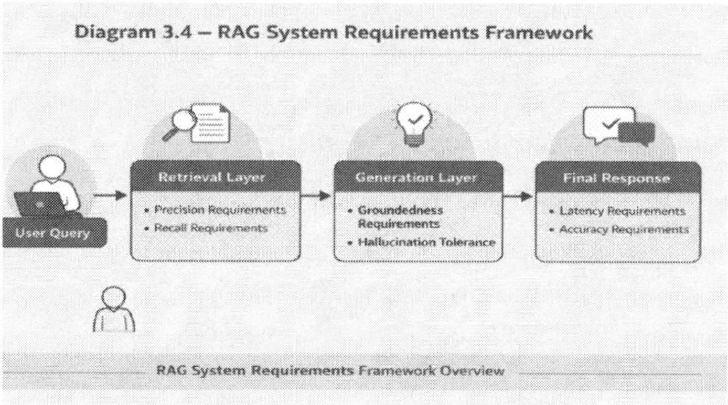

3.5.4 Context Window and Prompt Engineering Constraints

Every LLM has a maximum context window — the number of tokens it can process in a single request. Requirements should specify the minimum context window the system must support, the maximum prompt template length (system instructions, examples, and retrieved context combined), and the maximum response length. These three numbers determine whether a given model is architecturally suitable for the use case before any evaluation is conducted.

For RAG systems, the context window requirement interacts with retrieval design: the amount of retrieved context that can be included in a prompt is bounded by the context window minus the system instructions and the user query. A model with a 4,096-token context window leaves little room for retrieved documents, since system instructions occupy 1,500 tokens and user queries another

200 tokens. This constraint must be resolved in requirements, not discovered during prompt engineering.

3.6 Non-Functional Requirements

Non-functional requirements (NFRs) define the operational characteristics of a system: how it performs, how reliably it operates, how it scales, and how it recovers from failure. AI systems have a standard set of NFRs that overlap with those of conventional software systems, plus a set specific to machine learning workloads. Both must be captured in the requirements document.

3.6.1 Scalability and Throughput

Scalability requirements specify how the system must behave as request volume increases. A useful format: "The system must maintain p95 latency under 800 milliseconds when scaling from 100 to 1,000 concurrent requests, with linear cost scaling and no manual intervention required." This requirement captures three things: the latency constraint, the scale range over which the constraint applies, and the operational cost of scaling.

For model inference specifically, scalability often requires specifying whether the system uses auto-scaling, the warm-up time for new instances, and the behavior during scale-up lag (queue requests, return errors, or fail gracefully to a fallback system). These decisions should be made in requirements, not discovered at incident postmortems.

3.6.2 Availability and Recovery

Availability requirements for AI inference services follow the same patterns as conventional API services: target uptime (99.9% is three nines, approximately 8.7 hours of downtime per year), maximum tolerable downtime per incident, recovery time objective (RTO), and recovery point objective (RPO). For AI services, two additional requirements are relevant: the behavior during model provider outages (if the system depends on an external API) and the behavior when model performance degrades below a threshold (whether to fail open, fail closed, or fall back to a simpler system).

3.6.3 Cost Caps and Budget Triggers

Cost per inference was discussed above; it belongs in the NFRs alongside a cost cap — the maximum total monthly or annual spend the system is authorized to incur —, and budget triggers that mandate re-evaluation when usage approaches the cap. These are not aspirational targets; they are hard requirements with associated escalation paths. A system that can silently run up a $200,000 cloud bill is an incident waiting to happen.

For LLM systems with variable prompt length, cost estimation requires a model of the expected prompt-length distribution. Build that model from a representative sample of production traffic (or simulated traffic) before committing to a cost cap. A cap derived from average prompt length will be materially wrong for any workload with a long-tail distribution of complex queries.

3.7 Budgeting for Annotation and Human Review

The annotation budget is among the most underestimated line items in AI product plans, and the manager who does not establish it explicitly will discover it as an overrun. Annotation serves two functions in AI development: producing labeled training data and producing labeled evaluation data. Both are essential; both have distinct quality requirements; and neither should be treated as an afterthought.

Training data annotation budgets are driven by three variables: the number of examples required (which depends on model architecture, problem complexity, and desired accuracy), the cost per example (which depends on whether the annotation requires domain expertise and whether it is done internally or by a third-party labeling vendor), and the quality assurance overhead (which includes inter-annotator agreement measurement, arbitration of disagreements, and periodic audits of labeler quality). A rough rule of thumb for supervised classification tasks: budget 15 to 20 percent above the raw annotation cost for quality assurance and rework.

Evaluation data annotation is held to a higher quality standard than training data because errors in evaluation data directly corrupt the metrics used to make go/no-go decisions. For evaluation, use the most expert available annotators, enforce higher inter-annotator agreement thresholds, and maintain a chain of custody for the evaluation set to prevent inadvertent contamination by training data.

3.7.1 Ongoing Human Review

Beyond initial labeling, many production AI systems require ongoing human review: content moderation, quality sampling, edge case collection, and human-in-the-loop escalation handling. Budget for this work separately from the initial annotation. The review burden often increases during the first months of production as the system encounters distribution shift — inputs that differ systematically from training data. Plan for a review surge capacity of two to three times the steady-state review volume during the first ninety days of production.

Human review also generates a data asset: a stream of labeled examples drawn from production traffic. Capturing this asset requires review tooling that records the reviewer's decision, the time taken, the reviewer's identity, and any confidence score. That tooling investment is part of the annotation budget, not a separate infrastructure cost. The return on the investment is a continuously growing, production-realistic evaluation set that makes future re-evaluations cheaper and more accurate.

3.8 Stakeholder Negotiation Tactics

Getting engineering, data, and business stakeholders to agree on measurable requirements is a negotiation, and like all negotiations, it proceeds more smoothly with structure and preparation. The following tactics have proven reliable across a range of organizational contexts.

3.8.1 Lead with Business Consequences, Not Model Metrics

Data scientists and ML engineers are comfortable with precision, recall, and latency. Business stakeholders are not, and leading with these metrics in a mixed-stakeholder meeting produces disengagement or vague assent that conceals unresolved disagreement. Lead instead with the business consequence of model failure: "If the model makes a factual error on a billing question, the customer may take action based on that error, creating a customer service escalation and, in some cases, a chargeback." Then translate: "We therefore need a maximum factual error rate of one per five hundred billing-related responses, which corresponds to a precision requirement of 99.8% on that topic category." The business consequence makes the metric legible.

3.8.2 Use Ranges, Not Point Targets

Point targets create adversarial dynamics. If the requirement is "95% precision," then 94.8% is a failure and 95.1% is a pass — a distinction without business meaning. Ranges acknowledge the uncertainty inherent in model evaluation and shift the conversation from compliance to judgment. A requirement of "92 to 96% precision, with 94% as the design target and 91% as the absolute floor below which the product will not launch" creates a negotiation space that aligns incentives. Engineering knows the minimum that will be accepted; the business knows the aspirational target; and everyone has agreed on the threshold below which the product does not ship.

3.8.3 Create a Requirements Sign-Off Ceremony

Requirements that are discussed but never formally signed off tend to be quietly revised. Create a brief sign-off ceremony: distribute the requirements document two business days before a scheduled review meeting, hold the meeting to resolve open questions, and collect signatures (physical or digital) on the final document. This is not bureaucracy for its own sake. It is a social contract that changes the conversation from "I thought we all agreed" to "we agreed, and here is the document."

The sign-off should include representatives from the business (accountable for the outcome metrics), engineering (accountable for technical feasibility and latency/cost NFRs), data (accountable for training data, evaluation methodology, and achievable accuracy ranges), legal or compliance (accountable for privacy and regulatory constraints), and the product owner (accountable for the requirements document itself).

3.8.4 Handling the "We'll Know It When We See It" Objection

Some business stakeholders resist numerical requirements because quality is qualitative and cannot be reduced to a number. This objection, though common, is operationally untenable — it means the launch decision will be made on gut feel, and the post-launch dispute will be conducted in the same currency. The response is to offer a structured elicitation process: "Let's look at fifty examples together. You tell me which ones are acceptable and which are not. I'll build the requirement from your judgments rather than asking you to generate it from first principles." This approach works because it makes the abstraction concrete and demonstrates that the manager is not trying to win an argument but to capture what the stakeholder already implicitly knows.

3.9 The PRD Addendum Template

Every AI product initiative should carry a PRD addendum that supplements the standard product requirements document with AI-specific requirements. The addendum is not a technical specification — it is a business-level requirements contract written in language that all stakeholders can read and sign. The following checklist describes the essential components.

3.9.1 Core Components

The addendum should open with a business outcome statement in one sentence: what measurable business metric the product is expected to change, and within what timeframe. This is the anchor against which all technical requirements can be evaluated.

The model performance section should specify primary metrics (precision/recall/F1/accuracy as appropriate), target values and acceptable ranges, baseline comparators, the evaluation data set used to measure compliance (including how it was constructed and when it will be refreshed), and the acceptable failure mode taxonomy with rate thresholds.

The operational requirements section should specify latency targets by percentile, throughput targets, availability targets, RTO/RPO, cost per inference at the expected volume, and a cost cap with an escalation trigger.

The data requirements section should specify the training data volume, annotation vendor or internal team, inter-annotator agreement threshold, evaluation data volume, provenance separation from training data, data retention policy, and any data residency constraints.

For LLM and RAG systems, add a generative AI section covering hallucination tolerance, grounding requirements, context window minimum, privacy handling (zero-retention, data residency), prompt template governance (who owns and versions the system prompt), and the evaluation cadence for ongoing quality monitoring.

Diagram 3.5 – PRD Addendum Checklist Structure

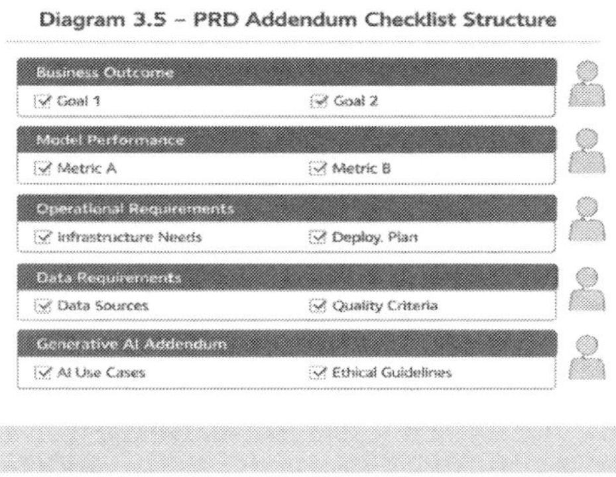

3.9.2 Living Document Governance

The PRD addendum is a living document. Requirements evolve as the team learns more about the data distribution, the user population, and the technical constraints. Establish a change control process: proposed changes to requirements must be submitted in writing, reviewed by all original signatories, and approved before taking effect. Changes made verbally in sprint planning meetings do not constitute approved requirement changes. This is a high bar, and it is intentional.

Track the addendum's version history alongside the product's version history. When a model is retrained or a prompt is changed, the associated requirements version should be recorded. This linkage makes incident investigation faster — when a model degrades in production, the first question is whether the requirements changed, then whether the data changed, then whether the architecture changed.

3.10 Vignette: Precise Requirements Save Weeks of Rework

A regional bank's fraud operations team was preparing to deploy a machine-learning model to replace a legacy rules-based fraud-screening system. The legacy system had a known false-positive rate of 22% — flagging one in five legitimate transactions for manual review — and the business goal was to reduce manual review volume while maintaining fraud catch rates.

In the first requirements meeting, engineering proposed optimizing for overall accuracy. The fraud operations lead pushed back: overall accuracy was misleading in a class-imbalanced dataset where 99.7% of transactions were legitimate. She wanted the recall on the fraud class to be no lower than the legacy system's 94%, the false positive rate to be no higher than 15%, and the latency to be under 200 *milliseconds to avoid blocking real-time transactions. She also specified a critical failure category: any fraudulent transaction above $10,000 that the model failed to flag would trigger an immediate incident review regardless of overall recall performance.*

This specificity felt excessive to some members of the team. It added an afternoon of requirements work and produced a two-page addendum. Two months later, during pre-production validation, the first model candidate achieved 96% recall on the fraud class but 19% false positive rate — within range on one metric, outside range on the other. Without the written requirements, there would have been pressure to launch and monitor. With the requirements in place, the conversation was unambiguous: the false positive threshold was 15%, the model was at 19%, and it was not ready to ship. The team spent three additional weeks adjusting the decision threshold and resampling the training data. The final model launched with 96.2% recall and a 13.8% false positive rate — within the window on both counts.

The fraud operations lead later noted that the requirements conversation had also raised a disagreement about the $10,000 critical-transaction rule, which would have caused a major dispute after launch. Engineering had not known about the threshold at all. Surfacing it in requirements costs an afternoon. Discovering it post-launch would have cost considerably more.

3.11 Manager's Checklist

Before moving from requirements to development, confirm each of the following items has been addressed and documented in the PRD addendum.

- Business outcome is stated in one sentence with a measurable metric and a timeframe.
- Business goal has been translated through an operational metric to a model behavior requirement.
- Primary model performance metrics (precision, recall, F1, accuracy) are specified with target values, acceptable ranges, and absolute floor thresholds.
- A baseline comparator is defined: the performance of the current process, a simple rules-based system, or a prior model version.
- Latency requirements are specified by percentile (p50, p95, p99) and by use case (synchronous vs. asynchronous).
- Throughput requirements are specified at expected production volume and at 2x expected volume for surge capacity planning.
- Cost per inference is specified at production volume, with a monthly cost cap and an escalation trigger.
- An error taxonomy has been developed with at least three levels (critical, significant, minor) and associated rate thresholds.
- Edge case categories are documented with handling requirements (support, detect and route, or reject with error).

- For LLM/RAG systems: hallucination tolerance, grounding requirements, context window minimum, and evaluation cadence are specified.
- Privacy and data handling requirements are specified: zero-retention policy, data residency constraints, and regulated data categories.
- Annotation budget includes training data volume, cost per example, quality assurance overhead, and evaluation data requirements.
- Ongoing human review budget is specified with surge capacity allowance for the first ninety days of production.
- All stakeholders (business, engineering, data, legal/compliance, product owner) have reviewed and signed the PRD addendum.
- A change control process for requirements is defined and communicated to the full team.

3.12 Takeaway

The requirements gap is where AI projects most consistently lose weeks and goodwill. A team that begins development without testable requirements will spend those weeks discovering, through expensive iteration, what should have been agreed upon in an afternoon. The manager's contribution to this work is not technical — it is the willingness to insist on precision, facilitate the stakeholder conversation, and hold the process to a written standard that can be audited later.

Measurable requirements serve three functions simultaneously. They give engineering a clear target, which accelerates development and eliminates the wasted cycles of building toward a moving or ambiguous goal. They give the business a basis for evaluating success, which transforms the launch conversation from a judgment call into a compliance check. And they create a shared record of what was agreed, transforming the post-incident conversation from blame assignment to root cause analysis.

Precision in requirements is not pessimism about the model or the team. It is operational clarity — the foundation on which everything else in the AI product lifecycle depends. Chapter 4 builds on that foundation by addressing the next major decision: selecting the model architecture that is best suited to the requirements just defined.

4 Model Selection and Architecture Tradeoffs

Neha Krishnamurthy, a senior product manager at a mid-size professional services firm, has just received sign-off to build an internal knowledge assistant. The assistant will answer questions drawn from thousands of policy documents, project retrospectives, and client engagement notes. Her engineering lead has placed three options on the table: plug into a frontier vendor API, build a retrieval-augmented generation (RAG) pipeline on top of an open-source language model, or fine-tune a smaller model on the firm's document corpus. Each path has a different story to tell. The API route could be live in a week, but legal flags data residency concerns because inference logs pass through a third-party cloud. The fine-tuning route promises deep accuracy but demands six weeks of labeling effort and a GPU budget that has not been approved. The RAG pipeline sits between them—moderate cost, reasonable accuracy, and data that never leaves the firm's Azure tenant. Neha has seventy-two hours to make a recommendation to the CTO.

This chapter is for every manager who has been in Neha's seat. It gives you a structured framework for making the architecture decision confidently, without needing to be an ML engineer. It explains the tradeoffs in plain language, provides the evaluation criteria your procurement and legal teams will need, and ends with a Build vs. Buy matrix you can adapt for your own organization.

4.1 Why It Matters

Architecture decisions are among the most expensive to reverse. A team that spends four months building on a vendor API only to discover a data-residency conflict faces an eight-to-twelve-week rebuild, a frustrated stakeholder, and a product timeline that has slipped by half a year. A team that jumps to fine-tuning before validating a simpler approach wastes annotation budget that could have funded two additional product hires. Getting the decision right early is not merely a technical preference—it is a business imperative.

Managers often defer architecture choices to engineering because the vocabulary feels unfamiliar. But the core tradeoffs—cost, latency, accuracy, risk, and operational burden—are squarely within a manager's domain. Engineers can model the infrastructure; only a manager can weigh regulatory constraints against roadmap pressure, or decide how much latency users will tolerate in exchange for a lower monthly compute bill.

The sections that follow give you a repeatable decision-making process. Each architecture option is described in terms of its cost, output, and requirements for your team. The goal is operational clarity: when you leave this chapter, you will know which questions to ask, which signals to watch for, and how to document your reasoning for leadership and for the engineers who will implement your decision.

4.2 The Model Selection Decision Framework

Before comparing architectures, lock down four dimensions. These are the axes on which every architecture option will be evaluated, and any decision made without explicit answers to them is likely to be reversed.

4.2.1 Dimension 1: Task Type and Accuracy Floor

What does the model need to do, and what is the minimum acceptable accuracy? A customer-facing chatbot that hallucinates one in ten responses has a different accuracy floor from an internal search tool where users can tolerate an occasional irrelevant result. Generative tasks (summarization, drafting, Q&A) have different requirements from classification or extraction tasks. Write down the task type and the error rate you can accept before any vendor conversation begins.

4.2.2 Dimension 2: Latency Tolerance

What is the maximum acceptable response time, and how does that vary by interaction pattern? A real-time assistant embedded in a customer portal may need to respond within 800 milliseconds. An overnight batch summarization workflow can tolerate 5 seconds per document. Latency tolerance is the single biggest variable separating which architectures are technically feasible. Self-hosted open-source models on underspecified hardware can be slower than a well-provisioned vendor API. Pin this number before you talk to engineers.

4.2.3 Dimension 3: Cost Sensitivity

What is the expected query volume, and at what cost per query does the product become economically unviable? At low volumes—say, five hundred queries a day—vendor API costs are negligible. At 50,000 queries a day, they can exceed $20,000 per month, at which point a self-hosted model pays for its upfront cost within 90 days. Run the unit economics for both ends of your expected volume range before treating cost as a secondary concern.

4.2.4 Dimension 4: Compliance and Data Residency

Which data classification does the content carry—public, internal, confidential, or regulated? Regulated data (PII, PHI, financial records) almost always prohibits sending content to a third-party inference endpoint without explicit contractual protections. Internal and confidential data require, at a minimum, a data processing agreement with the vendor. Establishing the data classification is a prerequisite, not an afterthought.

Diagram 4.1 – Model Selection Decision Framework

Diagram 4.1 – Model Selection Decision Framework

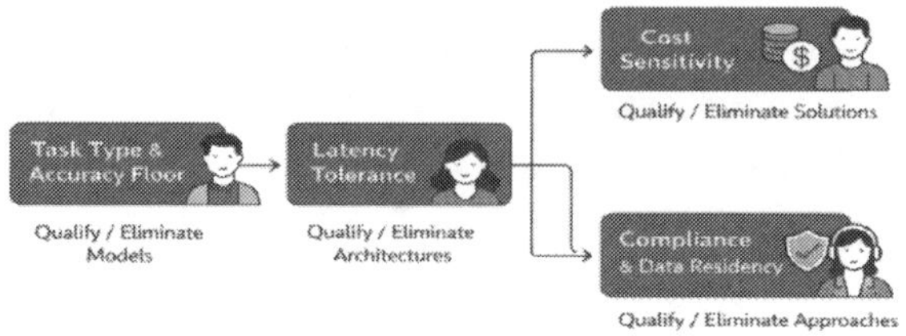

Evaluating Key Dimensions for Model Selection

4.3 Architecture Comparison: API, RAG, Fine-Tune, and Hybrid

There are five architectural patterns available to most product teams. They are not mutually exclusive, and the most capable production systems often combine elements from multiple systems. But for planning purposes, treat them as distinct paths with distinct cost, latency, and capability profiles.

4.3.1 Vendor API

A vendor API means your application calls a hosted language model—OpenAI, Anthropic, Google, Cohere, or similar—and pays per token consumed. Setup is minimal: an API key, a prompt, and a few lines of code. For general-purpose tasks (summarization, translation, Q&A on short documents), frontier models accessed via API deliver high accuracy with low engineering overhead. The tradeoffs are data sovereignty (all inputs and outputs pass through the vendor's infrastructure), cost at scale (token pricing grows linearly with volume), and model stability (the vendor controls when and how the model is updated).

Vendor APIs are the right starting point for exploratory work and proofs of concept. They are wrong for regulated data, high-volume, cost-sensitive workloads, and any use case that requires a reproducible, versioned model endpoint.

4.3.2 Retrieval-Augmented Generation (RAG)

RAG separates the knowledge store from the language model. Your documents are chunked, embedded as vectors, and stored in a vector database. At inference time, the user's query is embedded and used to retrieve the top-k most relevant document chunks, which are then passed to the language model as context. The model generates a response grounded in those retrieved chunks rather than in its pre-training knowledge alone.

RAG excels when the knowledge base is large, frequently updated, or proprietary. It dramatically reduces hallucination on in-domain questions because the model is always working from retrieved evidence. The architecture keeps data in your environment if both the vector store and the inference endpoint are self-hosted or within your cloud tenant. The downsides are retrieval latency (an extra 100–500 ms per query), chunk-quality sensitivity (poorly chunked documents degrade retrieval precision), and the need to maintain embedding freshness as documents change.

4.3.3 Fine-Tuned Open-Source Model

Fine-tuning takes a pre-trained open-source model—Llama, Mistral, Falcon, or similar—and continues training it on your domain-specific data. The result is a model that has internalized your vocabulary, your document formats, and your task patterns. Fine-tuning is appropriate when the task is narrow, labeled training data are available, and the volume economics justify the upfront investment.

The cost of fine-tuning is front-loaded: GPU compute, annotation labor, and evaluation rigor. Once deployed, the marginal cost per query is low because you own the weights and the inference infrastructure. The operational burden is high: you are now responsible for the model as a software artifact—versioning, monitoring, retraining when performance drifts, and patching security vulnerabilities in the serving stack. Most product teams underestimate this ongoing cost.

4.3.4 Hybrid: RAG Combined with Fine-Tuning

Hybrid architectures use a fine-tuned model as the reasoning engine and RAG as the knowledge retrieval layer. The fine-tuned model is trained to follow a specific instruction format and to cite retrieved passages faithfully; the RAG layer ensures the model has access to current, in-domain documents at inference time. This combination achieves the highest accuracy ceiling for complex domain tasks, but also the highest total cost and the greatest operational complexity.

Hybrid is appropriate for mature product teams that have first validated simpler architectures and found them insufficient. It is not a starting point—it is an endpoint for products that have earned their complexity through a proven user need.

Architecture	Cost Profile	Latency	Accuracy Ceiling	Ops Complexity
Vendor API	Pay-per-token; predictable at low volume, expensive at scale	Low–Medium (100–800ms)	High for general tasks; limited for domain-specific	Low
RAG Pipeline	Moderate; embedding + retrieval + LLM call	Medium (300–1500ms)	High for knowledge-grounded tasks	Medium
Fine-Tuned Open-Source	High upfront (GPU, labeling); low marginal cost	Low (self-hosted)	Very high for a narrow domain	High
Hybrid (RAG + Fine-Tune)	High upfront; variable runtime	Medium–High	Highest for complex domains	Very High
Custom Model (Scratch)	Prohibitive for most PMs	Variable	Best in class if resourced	Extreme

Table 4.1 – Architecture Comparison Summary. Use this table as a quick reference when presenting options to leadership or engineering partners.

Diagram 4.2 – Architecture Decision Tree: Accuracy vs Latency vs Cost vs Interpretability

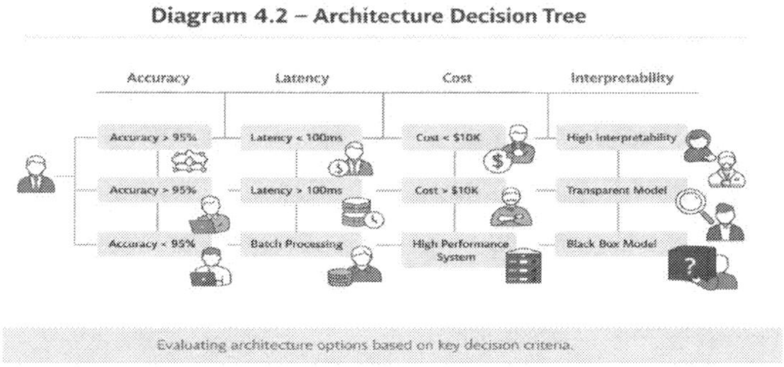

4.4 The Decision Tree: Accuracy vs Latency vs Cost vs Interpretability

No architecture is optimal on every dimension. The decision tree below provides a sequential evaluation logic. Work through it in order; the first failing gate eliminates one or more architecture options.

4.4.1 Gate 1: Interpretability Required?

If your use case requires the model to produce auditable, traceable reasoning—for regulatory compliance, legal review, or high-stakes clinical decisions—eliminate any black-box architecture. RAG with citation attribution is the most interpretable general-purpose option because outputs are grounded in retrievable source documents. Fine-tuned generative models are less interpretable without retrieval.

4.4.2 Gate 2: Data Residency Constraint?

If the answer is yes, vendor APIs that process data in third-party clouds are eliminated unless the vendor offers a private deployment option (e.g., Azure OpenAI Service within your own Azure subscription, or a dedicated inference endpoint). Self-hosted open-source models—fine-tuned or used as the backbone of a RAG pipeline—satisfy residency requirements by default.

4.4.3 Gate 3: Latency Requirement Under 500 ms?

If yes, a RAG pipeline with a remote vector store and a vendor inference endpoint may be too slow for synchronous, real-time use cases. Self-hosted models with a local vector store can achieve sub-200 ms retrieval plus inference. Vendor APIs alone (without retrieval) are typically the fastest option when the context window is sufficient for the task.

4.4.4 Gate 4: Volume Above 10,000 Queries per Day?

At this volume, the economic case for self-hosting a model—either fine-tuned or as a RAG backbone—typically becomes compelling within six to twelve months. Below this threshold, vendor API costs are manageable, and the reduced operational complexity is worth the premium. Run the break-even analysis with your finance partner before making a recommendation based solely on cost.

4.4.5 Gate 5: Narrow Domain with Labeled Data Available?

If you have a narrow, well-defined task and at least 500 to 1,000 labeled examples, fine-tuning becomes viable. Without labeled data, RAG is almost always the correct architecture for domain-specific accuracy because it requires only document ingestion—no labels.

Diagram 4.3 – Cost-Sensitivity Analysis: Volume vs. Monthly Spend by Architecture

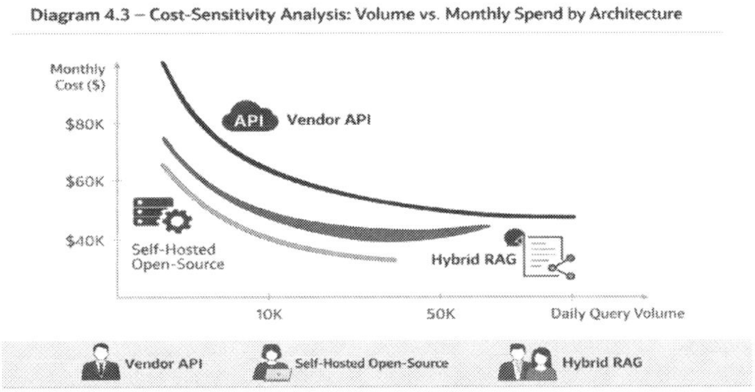

4.5 Cost-Sensitivity Analysis for Early POCs

Cost awareness at the proof-of-concept stage prevents unpleasant surprises at the production scale. Managers who approve a POC budget without modeling the production cost curve often find themselves defending a three-times budget overrun six months later. The following checklist structures cost awareness into the POC phase.

- **Estimate query volume at P50 and P99.** Model costs at the median and at the ninety-ninth percentile of expected volume, not just at the average. Traffic spikes are the rule, not the exception.
- **Calculate cost per query in three scenarios:** low volume (under 1,000/day), medium volume (10,000/day), and high volume (100,000/day). Identify which scenario triggers a build decision.
- **Include embedding costs in RAG architecture estimates.** Vector embedding via a vendor API adds token cost that is often overlooked in initial budgets.
- **Account for retrieval infrastructure.** Vector database hosting (Pinecone, Weaviate, pgvector) carries its own cost at scale. Include storage and query costs.
- **Amortize fine-tuning costs over the model's expected lifetime.** A fine-tuning run that costs twenty thousand dollars amortizes to less than two cents per query over a twelve-month production run at moderate volume.
- **Model retraining frequency.** Fine-tuned models require periodic retraining as domain knowledge evolves. Budget for a retraining cycle every three to six months.
- **Include human review labor.** Any architecture that requires human review of model outputs must include reviewer labor in the unit economics.
- **Build a cost dashboard before launch.** Token spend, retrieval latency, and compute cost should be visible to the PM in real time, not discovered in a monthly billing statement.

4.6 Vendor Evaluation Criteria

When a vendor API is the selected architecture—or when a vendor supplies a component of a RAG or hybrid system—procurement and legal teams will require structured evaluation inputs. The criteria below frame the questions you need answered before a vendor is approved.

4.6.1 Data Residency and Processing

Establish exactly where data is processed and stored. Third-party inference APIs process your prompt and return a completion; in doing so, they receive your input data, which may include confidential or regulated content. Ask for the vendor's data processing agreement (DPA), identify which region hosts inference and logging, and confirm whether inputs are used for model training by default—and how to opt out.

4.6.2 Service Level Agreements

A vendor's advertised uptime percentage is not a service level agreement. An SLA must specify the uptime commitment, the measurement window (per month is standard), the remediation mechanism (credits, not refunds, are typical), and the resolution time commitment for incidents. A 99.9% uptime SLA permits approximately 8.7 hours of downtime per year; 99.5% permits 43.8 hours. Understand what interruption your users can absorb before accepting a vendor SLA.

4.6.3 Model Update Cadence and Versioning

Vendor models change. OpenAI, Anthropic, and Google all update their model endpoints, and those updates can alter output behavior in ways that break prompt-engineered applications. Ask whether the vendor offers stable, versioned endpoints that do not change without notice, and what the minimum deprecation notice period is. A vendor that cannot commit to versioned stability creates ongoing regression risk that your QA process must absorb.

4.6.4 Observability and Log Export

Monitoring a vendor-hosted model requires access to request and response logs, token usage, and latency data. Ask whether the vendor provides a log export mechanism, what the retention period is, and whether latency percentiles are exposed via a dashboard or API. Without observability, you cannot detect performance degradation, cost anomalies, or compliance incidents.

Criterion	What to Ask	Red Flags
Data Residency	Where is inference executed? Which region stores input/output logs?	"Data may be used for training" without explicit opt-out.
SLA / Uptime	What is the guaranteed uptime? What are	SLA < 99.5%; remediation is credit-

	credit/remediation terms?	only with no SLA for resolution
Model Update Cadence	How often does the underlying model change? Is there a stable endpoint?	Vendor updates model silently; no versioned endpoint offered
Observability Hooks	Can you export request/response logs? Are token usage and latency visible?	No log export; no per-request latency visibility
Contractual Exit Rights	What is the notice period? Is there an SLA for data deletion upon termination?	Auto-renew with 90-day notice; no data deletion commitment

Table 4.2 – Vendor Evaluation Criteria. Use this table as an input to RFP responses and procurement review.

4.7 Integration Risk Assessment

Architecture decisions do not exist in isolation. They integrate with existing systems—authentication, logging, user-facing products, data pipelines—and each integration point carries risk. A manager who approves an architecture without an integration risk assessment often discovers those risks during implementation, when reversal is most expensive.

4.7.1 Latency Budget

Map the full request path from user input to response delivery. Each component in the path—authentication, routing, vector retrieval, inference, response parsing, rendering—adds latency. Sum the P99 latency for each component and compare it to your user-facing latency requirement. If the sum exceeds your budget, you must either optimize a component or change the architecture. Do this analysis on paper before implementation begins.

4.7.2 Fallback Paths

What happens when the primary model endpoint is unavailable? Every production AI system needs a fallback: a simpler model, a deterministic keyword search, a cached response, or a graceful degradation that informs the user that the system is temporarily unavailable. Fallback paths must be designed, implemented, and tested—not assumed to be someone else's responsibility. Define the fallback behavior as part of the architecture decision.

4.7.3 Vendor Lock-in Mitigation

Vendor lock-in risk is highest when the application logic is tightly coupled to a specific model's output format, function-calling schema, or proprietary API idioms. Mitigate this by abstracting the model interface behind a thin service layer that can swap the underlying model without changing downstream application code. This abstraction adds a small engineering overhead upfront and can save weeks of rework if the vendor changes pricing, deprecates a model, or introduces a breaking API change.

4.7.4 Authentication and Access Control

Vendor API keys are high-value credentials. Exposure of an API key grants the holder unrestricted access to your account—including the ability to run queries at your expense and potentially access logged data. Establish secrets management from day one: store keys in a vault (Azure Key Vault, AWS Secrets Manager, HashiCorp Vault), rotate them on a schedule, and audit key usage. Never embed API keys in application code or environment files committed to source control.

Diagram 4.4 – Integration Risk Map for AI Architecture Components

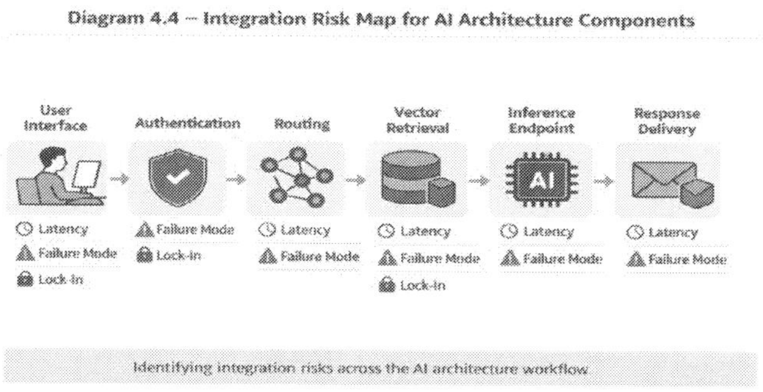

4.8 Build vs. Buy Decision Matrix

The build vs. buy decision is not binary. Most real-world AI architectures combine built components (custom data pipelines, fine-tuned model weights, proprietary evaluation harnesses) with bought components (managed vector databases, vendor inference APIs, observability platforms). The matrix below structures the decision by criterion.

Criterion	Build (Open-Source / Custom)	Buy (Vendor API / SaaS)
Data Residency	Full control; deploy in your environment	Risk: data leaves the boundary unless contracted
Time to First Value	Weeks to months	Hours to days
Customization Depth	Unlimited; full model access	Limited to prompt engineering and fine-tune APIs
Operational Burden	High: MLOps, infra, security, patching	Low: vendor manages stack
Cost at Scale	Low marginal cost after amortization	Grows linearly with volume; can be prohibitive
Vendor Lock-in Risk	None; own the weights	High: model changes, price hikes, deprecation

Team Capability Required	ML engineers, infra, annotation expertise	Product and prompt-engineering skills are sufficient
Best Fit	High-volume, regulated, or IP-sensitive workloads	Exploratory POCs, fast launches, generalist tasks

Table 4.3 – Build vs. Buy Decision Matrix. Score each criterion for your specific context and sum the "Build" vs. "Buy" signals to inform the recommendation.

When using this matrix, resist the urge to treat it as a mechanical scorecard. Two criteria—data residency and vendor lock-in—can override most other signals. A product with a compelling economic case for a vendor API should still build a self-hosted solution if the data classification prohibits third-party processing. The matrix is a structured starting point for the conversation, not a substitute for judgment.

4.9 Vignette: The Hybrid RAG Decision

Returning to Neha's knowledge assistant project: after working through the decision framework, she documents her analysis in a one-page architecture brief for the CTO.

The task type is open-domain Q&A over a heterogeneous document corpus of twelve thousand files. The accuracy floor is that fewer than one in twenty answers should be factually incorrect or unsupported by a source document—a 95% grounding requirement. The latency tolerance, set by a user experience benchmarking session, is

1.5 seconds for 90% of queries. The query volume in year one is projected at approximately 8,000 queries per day, growing to 25,000 by year two. The data classification is confidential—most documents contain client engagement details that must remain within the firm's Azure tenant.

The vendor API route is eliminated at Gate 2 due to data residency. The firm's DPA with the leading frontier model provider does not cover confidential client data for inference, and the vendor's private deployment option costs four times the standard API rate, pushing the cost over budget.

The pure fine-tuning route is eliminated at Gate 1: the corpus is updated weekly, and a fine-tuned model cannot incorporate new documents without retraining. The firm cannot accept a one-to-two-week lag between a new document being published and the assistant being able to answer it.

The RAG pipeline, deployed entirely within the firm's Azure tenant using Azure OpenAI Service (private endpoint) as the inference layer and Azure Cognitive Search as the vector store, satisfies all four gates: it respects data residency, achieves measured P90 latency of 1.2 seconds in load testing, costs approximately 0.008 dollars per query at the projected year-one volume, and can incorporate new documents within four hours of ingestion.

Neha documents a clear rollback path: if Azure OpenAI Service SLA breaches occur more than twice in a quarter, the team will evaluate a self-hosted Mistral deployment using the same retrieval layer. The architecture brief includes a decision log, cost model, and rollback criteria. The CTO approves it in one meeting.

The lesson is not that RAG is always the right answer. The lesson is that Neha made the decision systematically, documented the reasoning at each gate, and built in a rollback path before writing a line of code. That discipline—not the technology choice itself—is what a manager's architecture decision should look like.

Diagram 4.5 – Vignette: Hybrid RAG Architecture for the Knowledge Assistant

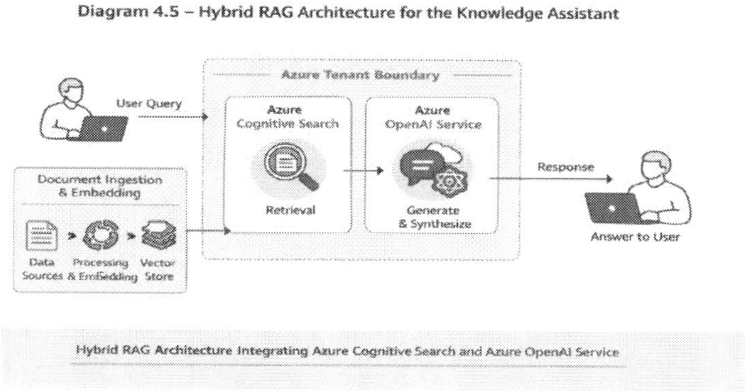

4.10 Manager's Checklist

- Define the four decision dimensions before any architecture discussion: task type, latency tolerance, cost sensitivity, and data classification.
- Run the decision tree gate sequence: interpretability, data residency, latency, volume, and labeled data availability.

- Require unit economics at three volume levels—low, medium, and high—before approving a vendor API for production use.
- Verify data residency compliance with legal before any POC that uses confidential or regulated data.
- Confirm vendor SLA terms, model versioning policy, and log export capability before signing a contract.
- Map the full request latency budget across all integration components before committing to a real-time use case.
- Define fallback behavior for every AI-powered feature before development begins.
- Abstract the model interface behind a service layer to reduce vendor lock-in from day one.
- Rotate and vault all vendor API keys; never embed credentials in source code.
- Document the architecture decision, including eliminated options and the reasoning for each gate outcome.
- Build in a rollback path with explicit trigger criteria before the architecture is approved.
- Establish a cost dashboard that is visible to the PM before the product goes live.
- Review the architecture decision at the ninety-day mark against actual query volume, latency, and cost data.

4.11 Takeaway

Architecture choices in AI products are decision-making events, not technical details to be delegated to others. The product's owner owns the architecture trade-offs that shape its cost, reliability, and compliance posture. The decision framework in this chapter—four evaluation dimensions, a five-gate decision tree, structured vendor evaluation criteria, and a build vs. buy matrix—gives you the vocabulary and the structure to lead that conversation rather than be a passenger in it.

The most common architecture failure mode is choosing the wrong technology. It is choosing the right technology for the wrong reasons: selecting a vendor API because it is fast to implement, without modeling the production cost curve; choosing fine-tuning because the accuracy ceiling is higher, without validating that labeled data and GPU budget are available; adopting a hybrid architecture because it sounds sophisticated, without earning that complexity through a validated, simpler path.

Discipline in the architecture decision phase compounds over the product's life. A well-documented decision log, a cost model reviewed at each growth milestone, a rollback path never invoked but always maintained, and a vendor evaluation process that respects data residency from the start—these practices are the foundation of AI products that scale without crisis. Own the tradeoff analysis. Document the reasoning. Build in the exit. That is the manager's job.

5 Data Strategy, Labeling, and Feature Stores

The alert came from a product manager on a Tuesday afternoon: the recommendation model had been surfacing irrelevant content for three weeks. Engagement rates had dropped by eighteen percent. Customer support tickets mentioning "bad suggestions" had tripled. When the engineering team dug into the logs, the culprit was not a model change—it was a seemingly routine pipeline migration. A data-engineering team had renamed a field in the user-event schema to comply with a new naming convention. The feature store had no validation gate, the model silently consumed null values in place of the renamed column, and no dashboard flagged the shift because the alerting thresholds were set on business metrics alone, not on upstream feature distributions. No one noticed for twenty-two days.

This chapter is about preventing that story from being yours. It covers how to design production-grade data pipelines, how to build quality gates that catch problems before they reach a model, how to structure labeling programs so that the data you invest in actually moves the needle, how to use feature stores to keep training and inference in sync, and how to budget annotation effort rationally. Done right, data strategy is not an infrastructure detail—it is the foundation of every product guarantee you make to your business stakeholders.

5.1 Why It Matters

Data is the product's fuel. A model trained on stale, mislabeled, or unrepresentative data will degrade silently. Unlike software bugs—which typically produce visible errors—data problems are insidious. The model continues to serve predictions. Response codes remain 200. Latency stays within SLA. The only signal is a slow, almost imperceptible erosion of business outcomes that most teams attribute to seasonal variation, shifting user behavior, or marketing spend rather than a broken data pipeline.

For a manager, the practical consequence is clear: without deliberate data governance, you have no reliable way to know whether your model is working as designed. You are flying blind over complex terrain. Governance does not mean bureaucracy—it means installing the instruments that let you see. The sections that follow treat data strategy as a product discipline rather than a technical afterthought. Each topic is presented through the lens of decisions you must make, risks you must manage, and actions your team must take.

The opportunity is equally significant. Teams that treat labeling as a strategic lever—not a cost to minimize—consistently outperform those that outsource it without governance, or skip it in favor of larger pre-trained models. Annotation quality compounds: a well-governed labeling program produces consistent, reusable training data, making every subsequent fine-tuning run cheaper and faster. The remainder of this chapter explains how to build that compounding advantage.

5.2 Designing the Production Data Pipeline

A production data pipeline is not a one-time ETL script. It is a living system that transforms raw source data—events, transactions, user records, third-party feeds—into model-ready features on a schedule that matches your inference latency requirements. Understanding this distinction matters for a manager because it changes where you invest, what you monitor, and how you structure team ownership.

5.2.1 From Source to Model-Ready Data: The Five Zones

Most mature data pipelines can be conceptualized as five logical zones, each with distinct ownership and quality concerns.

The ingestion zone collects raw data from operational systems, including clickstreams, API calls, database change logs, IoT sensors, and third-party data vendors. This zone should be append-only—data lands without transformation so that every downstream problem can be reproduced and debugged. The transformation zone normalizes, joins, and enriches raw records into clean, consistently typed datasets. The feature engineering zone computes model-specific signals from clean data: rolling aggregates, embeddings, categorical encodings, and interaction terms. The serving zone moves computed features into low-latency storage for real-time inference. The archival zone stores historical snapshots for retraining, auditing, and regulatory compliance.

Each zone boundary is a natural insertion point for quality gates (covered in the next section). Owning the boundary between zones means owning the contract: the schema, the latency SLA, the acceptable null rate, and the alert that fires when any of those contracts is violated.

Diagram 5.1 – Production Data Pipeline – Five-Zone Architecture

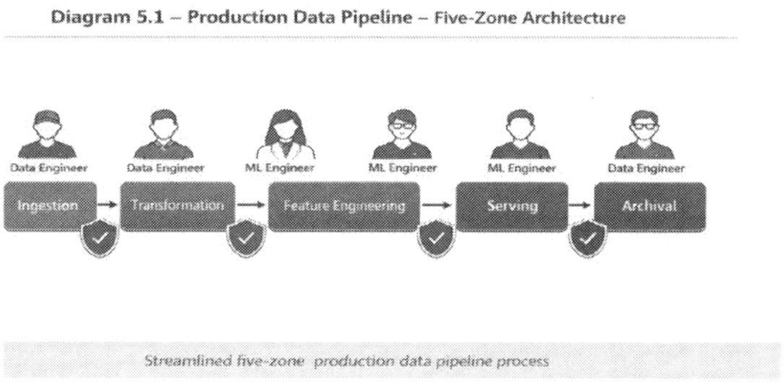

5.2.2 Pipeline Ownership and Handoff Contracts

One of the most common failure modes in AI product organizations is ambiguous pipeline ownership. When a feature breaks, everyone assumes it is someone else's domain. The fix is explicit handoff contracts between the data engineering team (which owns ingestion through transformation) and the ML engineering team (which owns feature engineering through serving).

A handoff contract should specify: the expected schema (column names, types, allowed values), the latency guarantee (how fresh the data must be at serving time), the acceptable null rate for each field, and the escalation path when the contract is violated. This contract should live in the same repository as the pipeline code so that it changes only through a reviewed pull request, not a verbal agreement in a sprint planning meeting.

As the product manager, your job is not to write the contract—it is to ensure the contract exists, is tested, and is visible to both teams. If you cannot point to a written pipeline contract during an incident review, you have a governance gap.

5.2.3 Incremental vs. Full-Refresh Pipelines

A persistent architectural decision is whether to run incremental pipelines—which process only new or changed records since the last run—or full-refresh pipelines, which reprocess all historical data on each execution. Incremental pipelines are faster and cheaper but harder to get right; late-arriving data and out-of-order events can silently corrupt aggregates. Full-refresh pipelines are simpler to reason about, but become computationally prohibitive at scale.

The practical manager's heuristic: start with full-refresh pipelines during the prototype stage, when correctness matters more than speed. Migrate to incremental pipelines only when you have automated tests that verify correctness and a monitoring plan that can detect late-data corruption. Never migrate to incremental pipelines as a cost-cutting measure without also investing in the testing infrastructure that makes incremental processing safe.

5.3 Quality Gates for Data Integrity

A quality gate is an automated check that runs at a pipeline boundary and halts or alerts when data does not meet a defined threshold. Quality gates are the difference between discovering a data problem in production after three weeks of silent degradation and discovering it in the pipeline within minutes of occurrence.

5.3.1 The Four Classes of Data Quality Checks

Schema checks verify that columns exist, have the expected data types, and fall within allowed categorical values. These are the lowest-cost checks to implement and the highest-value to run first—most pipeline breakages result in schema violations.

Completeness checks verify that null and missing-value rates are within acceptable bounds. A feature column that was 5% null last week and is now 40% null signals a broken upstream source, not normal variation. Each feature should have an explicitly configured null tolerance threshold, not a generic default.

Distribution checks verify that the statistical properties of each feature—mean, standard deviation, quartiles, or histogram bin counts—have not shifted beyond a configurable threshold. Distribution drift is the early warning signal for concept drift in the model; catching it in the pipeline, before it reaches inference, prevents the silent degradation scenario that opened this chapter.

Referential integrity checks verify that foreign key relationships hold—that user IDs in the event log exist in the user table, that product IDs in the recommendation feed match active catalog items. These checks are especially important in federated data architectures where different teams own multiple upstream systems.

Diagram 5.2 – Quality Gate Decision Tree – Pipeline Check Hierarchy

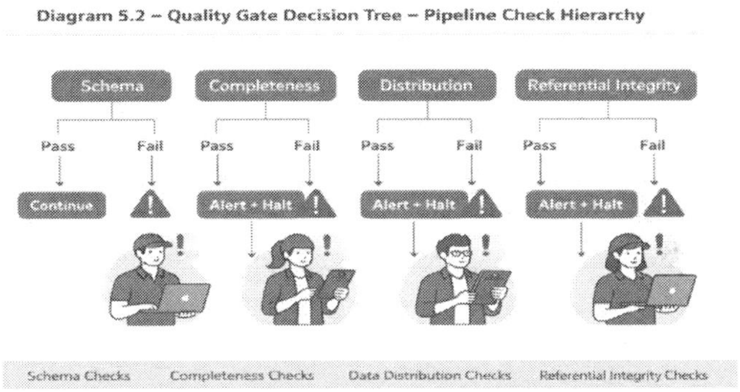

5.3.2 Configuring Thresholds and Alerts

Threshold configuration is where most teams make their first mistake: they pick round-number defaults (five percent null tolerance, ten percent distribution shift) without grounding them in historical baselines. The correct approach is to observe your pipeline for two to four weeks in a stable state, compute the natural variation bands for each metric, and set thresholds at a meaningful multiple of that natural variation—typically two standard deviations for warning alerts and three for hard stops.

Alert routing matters as much as alert configuration. A data quality alert that goes to a shared Slack channel and is never acknowledged is worse than no alert—it creates alert fatigue that causes engineers to mute the channel. Route schema failures to the data engineering on-call. Route distribution shift warnings to the ML engineering on-call. Route business-metric anomalies to the product manager. Clear routing produces clear ownership; clear ownership produces fast response times.

5.3.3 Automated Remediation vs. Human Escalation

Not every quality gate failure requires human intervention. Some classes of failure have well-understood remediation paths that can be automated: a late-arriving batch can trigger an automatic retry; a schema mismatch can trigger a schema-migration job if the change is additive. Define an automated remediation playbook for the most common failure modes, and reserve human escalation for failures that fall outside the playbook. This approach keeps your on-call burden manageable while ensuring that novel failures receive proper attention.

5.4 Labeling Workflows and Annotation Governance

Annotation is where data strategy meets product quality most directly. The labels your team produces are the ground truth against which every model evaluation is measured. Annotation governance is the set of processes, standards, and checks that ensure those labels are consistent, accurate, and representative.

5.4.1 Labeler Training and Qualification

Annotation quality starts with labeler preparation, not labeler selection. Even experienced annotators produce inconsistent labels when given ambiguous guidelines. A labeler training program should include: a written annotation guide with examples of each label class (including edge cases and deliberate hard negatives), a calibration exercise where new labelers annotate a pre-labeled gold set and must achieve a minimum agreement score before working on production data, and periodic recalibration sessions when guidelines change or when annotation statistics drift.

The annotation guide is a living document—it should be versioned and reviewed whenever model performance degrades in a way that suggests label inconsistency. Treat it with the same discipline you would apply to a product requirements document: every change should be reviewed, approved, and communicated to active labelers before it takes effect.

5.4.2 Inter-Annotator Agreement

Inter-annotator agreement (IAA) is the statistical measure of how consistently multiple annotators apply the same label to the same item. Cohen's Kappa is the standard metric for binary and categorical labels; Krippendorff's Alpha is preferred for ordinal or continuous ratings. As a manager, you do not need to compute these metrics yourself—you need to ensure that your annotation platform or your ML team reports them to you on a regular cadence and that you have defined acceptable thresholds.

A rule of thumb: Kappa below 0.60 indicates substantial labeler disagreement and should halt production labeling until the root cause is identified (usually ambiguous guidelines or undertrained labelers). A kappa between 0.60 and 0.80 is acceptable but warrants investigation. A kappa above 0.80 indicates strong agreement and is your target for high-stakes label categories.

Disagreement data is as valuable as agreement data. When two labelers consistently disagree on a particular item type, that disagreement is often a signal that the item type is genuinely ambiguous—which may indicate a product definition problem, not just a labeling problem. Surface these cases to your product and domain-expert stakeholders rather than resolving them solely through majority voting.

Diagram 5.3 – Annotation Governance Workflow – From Task Creation to Approved Label

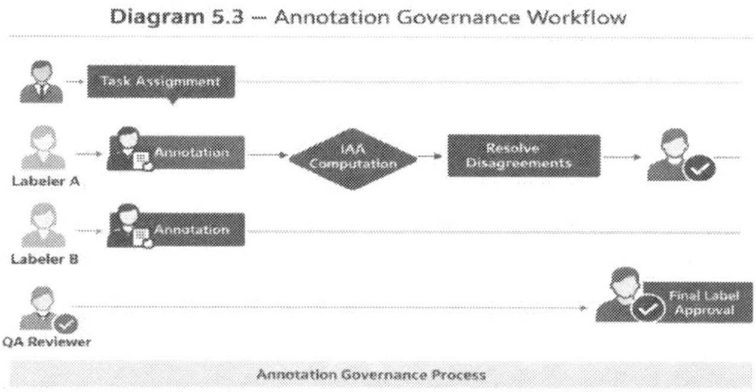

5.4.3 Quality Assurance in Annotation Programs

Three QA mechanisms are standard in mature annotation programs. Honeypot insertion places pre-labeled items of known correct answers into the regular annotation queue; an annotator's performance on honeypots is a real-time proxy for their overall accuracy. Consensus review routes a sample of completed annotations to a second labeler or to a domain expert for independent verification; the agreement rate on this sample estimates the overall label accuracy of the batch. Audit trails log every annotation decision, revision, and reviewer comment, creating an immutable record that supports downstream model investigations and regulatory inquiries.

As a manager, the critical governance decision is what to do when QA reveals systematic labeler error. The wrong response is to correct the bad labels and move on quietly. The right response is to treat the error as a process failure: retrain or recalibrate the labeler, investigate whether other batches are affected, and update the annotation guide to prevent recurrence.

5.5 Feature Stores for Repeatability and Consistency

A feature store is a centralized platform for storing, serving, and sharing machine learning features. The business case for a feature store is straightforward. Without one, every team that builds a model recomputes the same features from scratch, often with subtle differences in logic that make their models incompatible. With one, features are computed once, validated once, and reused across models and teams.

5.5.1 The Training-Serving Skew Problem

Training-serving skew is the silent killer of production ML systems. It occurs when the features used to train a model are computed differently from the features used to serve predictions. This can happen for seemingly trivial reasons: a timestamp timezone conversion that behaves differently in batch and streaming, a join key that handles nulls differently in the offline training pipeline and the online serving layer, or a feature that was backfilled historically but is computed in real time at inference.

A feature store solves training-serving skew by enforcing a single feature definition. The same code that computes a feature for the training dataset also computes it at serving time. When that code changes, both training and serving change together—the feature store is the single source of truth.

5.5.2 Offline Store, Online Store, and Point-in-Time Correctness

Feature stores typically expose two storage tiers. The offline store (usually a columnar data warehouse or object store) holds the full historical record of feature values; it is queried during model training to retrieve the exact feature values that existed at the time of each training example. The online store (usually a low-latency key-value store) holds only the most recent feature values; it is queried at inference time to enrich a prediction request with the current feature state.

Point-in-time correctness is the property that training examples are assembled using only the feature values that would have been available at the time the label was generated. Violating this property—by using future feature values to predict past labels—produces training data with look-ahead bias, which inflates offline evaluation metrics while producing a model that cannot replicate that performance in production. Your ML team should be able to demonstrate that their training data assembly process is point-in-time correct; if they cannot, you have a data leakage risk that will surface as an offline-online performance gap.

Diagram 5.4 – Feature Store Architecture – Offline and Online Tiers

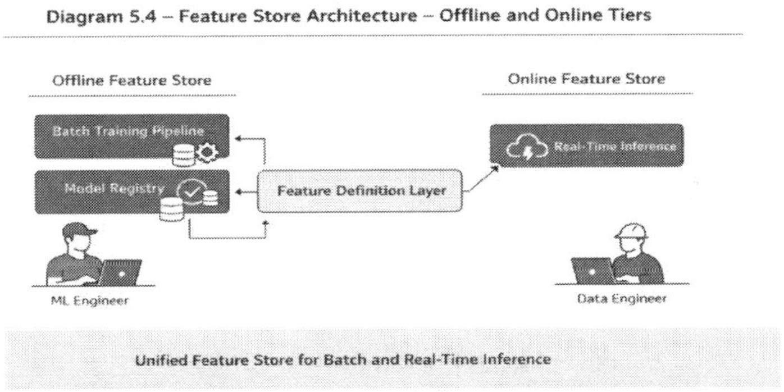

5.5.3 Governance and Feature Discovery

The governance value of a feature store is proportional to its level of documentation. A feature that exists in the store but has no description, no owner, and no lineage information is a liability, not an asset—it will be misused, misinterpreted, or silently deprecated. Require that every feature registered in your store has: a plain-language description of what the feature measures, the upstream source and transformation logic, the owning team and point of contact, the model or models that currently consume it, and the last validation date.

Feature discovery—the ability for a new team to search the feature store and find a feature that meets their needs—reduces duplicated computation and accelerates model development. A well-documented, searchable feature store is a force multiplier for your entire data science organization. A poorly documented feature store is a maintenance burden that no one trusts.

5.6 Budgeting Annotation Effort Against Model Improvement

Annotation is expensive. At scale, it is often the single largest line item in an AI product's operating budget. The manager's challenge is to allocate that budget to the labeling work that produces the most model improvement per dollar spent—and to resist the temptation to label as much data as possible without first asking what specific capability gap the labeling effort is meant to close.

5.6.1 The Diminishing Returns Curve

Every model has a learning curve that relates the number of labeled examples to model performance on a target metric. Early in a training data program, each labeled example produces a large marginal improvement—the model is learning basic patterns. As the dataset grows, additional examples produce smaller marginal improvements—the model has seen most of the easy cases. Eventually, additional labeling of randomly sampled data yields almost no improvement, and the only way to continue improving is to label strategically, focusing on the specific slices where the model currently fails.

As a manager, you should require your ML team to produce an empirical learning curve for your model before committing to a large labeling budget. A learning curve plots model performance against dataset size, using experiments at multiple dataset sizes to fit the curve. If the curve has already entered the flat region, random labeling is a poor investment; the team should instead pursue targeted labeling (covered in the next section) or investigate changes to the model architecture.

5.6.2 Cost per Annotation and Return on Investment

The cost per annotation varies widely depending on task complexity, the labeler profile (crowdsourced workers, domain experts, or clinicians, for example), and the quality-control overhead. A simple binary classification label from a crowdsourced platform might cost 10 cents per example; a complex clinical annotation requiring a licensed physician might cost \$50 or more. Knowing your cost per annotation is a prerequisite for any ROI conversation with leadership.

The return side of the equation is harder to quantify but essential to attempt. If your model's current precision is 0.72 and your business case projects that raising it to 0.80 would recover three million dollars annually in mislabeled orders, then you have a framework for deciding how much to spend on labeling. Work with your data science team to estimate the performance improvement that each labeling scenario—more random data, more targeted data, a specific slice—is likely to produce, and use that estimate to bound the ROI conversation.

5.7 Sampling Strategies for High-Impact Edge Cases

Random sampling from your production traffic is the baseline labeling strategy, but it is rarely the most efficient. Production traffic reflects the distribution of common cases; models typically handle common cases well after initial training. The cases that degrade performance, damage customer experience, and generate escalations are edge cases—rare, unusual, or systematically underrepresented inputs that the model has not learned to handle.

5.7.1 Failure-Driven Sampling

The highest-ROI labeling work targets the specific inputs where the model currently fails. Failure-driven sampling starts by identifying a set of model failures from production logs—low-confidence predictions, high-disagreement outputs, user corrections, or negative feedback signals—and routing those specific inputs to the labeling queue. Because each labeled example is a case the model got wrong, every annotation directly addresses a known weakness.

Failure-driven sampling requires a feedback loop between your production inference system and your labeling queue. Building this loop is a one-time engineering investment that pays dividends over the product's lifetime. If your team has not built this loop, it should be on your roadmap.

5.7.2 Uncertainty Sampling and Active Learning

Uncertainty sampling is the annotation-efficiency technique of selecting for labeling the examples about which the model is least confident—those where the predicted probability distribution is flattest or where an ensemble of models disagrees most. By labeling uncertain examples first, you provide the model with exactly the information it needs to reduce its uncertainty, compressing the learning curve and reducing the total number of labels required to reach a target performance level.

Active learning is the broader paradigm that uses model uncertainty to guide data collection in an iterative loop: label a batch, retrain the model, identify the new most uncertain examples, label them, and repeat. Active learning can reduce labeling costs by 30 to 50 percent compared to random sampling on many tasks. Still, it requires close coordination between the labeling and ML teams and a fast model retraining cycle to be practical.

5.7.3 Slice-Based Sampling for Demographic and Operational Coverage

Some sampling decisions are not about model uncertainty—they are about ensuring that specific slices of your input distribution are represented in your training data. A product that serves customers across multiple geographies, languages, age groups, or usage patterns must ensure that its training data is representative across all of those dimensions, not just in aggregate.

Slice-based sampling defines the critical dimensions of your product's input space. It allocates a portion of your labeling budget to each slice in proportion to its business importance, not its frequency in production traffic. A low-frequency but high-value slice—such as large enterprise customers making high-stakes decisions—may warrant a labeling investment that exceeds its raw traffic share. Document your slice definitions and their labeling budget allocations as part of your data strategy, so that coverage decisions are explicit and auditable.

Diagram 5.5 – Sampling Strategy Comparison – Random vs. Targeted vs. Active Learning

Diagram 5.5 - Sampling Strategy Comparison

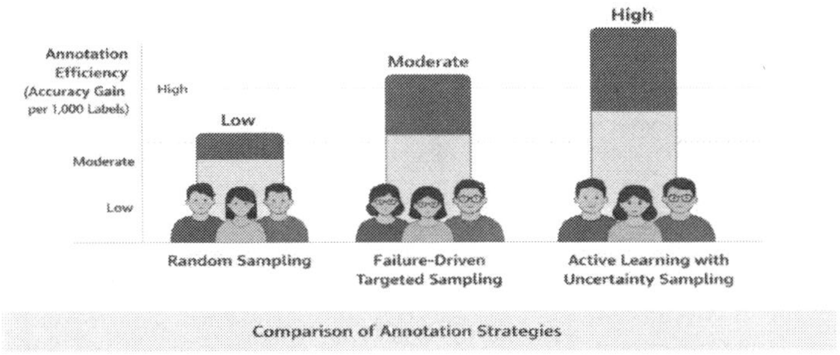

5.8 The Data Readiness Checklist

Before committing to a model development sprint, production deployment, or significant annotation investment, run every data initiative through the following checklist. This is designed to be completed in a single working session by the product manager, ML engineer, and data engineer jointly. Any item marked "No" or "Unknown" is a gap that must be resolved or explicitly accepted as a risk before the initiative proceeds.

5.8.1 Pipeline and Infrastructure

- All data sources identified and access permissions confirmed
- Handoff contracts documented for each pipeline zone boundary
- Schema checks, completeness checks, and distribution checks are deployed at each zone boundary
- Alert routing defined: each alert type has a named owner and an expected response time

- Automated remediation playbooks written for the three most common failure modes
- Historical data available in sufficient volume for the training window required
- Point-in-time correctness verified for training data assembly

5.8.2 Feature Store and Feature Quality

- All model features registered in the feature store with description, owner, and lineage
- Training-serving parity confirmed: no features computed differently in offline and online paths
- Feature freshness requirements documented and matched to pipeline latency SLAs
- No deprecated or undocumented features consumed by the production model

5.8.3 Labeling and Annotation

- Annotation guide is current, versioned, and reviewed by domain experts
- Labeler calibration scores on the gold set meet or exceed the minimum threshold
- Inter-annotator agreement reported and within an acceptable range
- QA mechanism in place: honeypots, consensus review, or audit trail
- Labeling budget approved and ROI estimate documented
- Sampling strategy defined: not purely random if targeted sampling is more efficient

- Slice coverage audited: all critical input slices are represented in training data

5.9 Vignette: A Focused Labeling Sprint That Moved the Needle

A product team managing an e-commerce search relevance model had been struggling with flat performance on apparel queries. The model's overall MRR (mean reciprocal rank) was acceptable at 0.74. Still, qualitative reviews consistently flagged that apparel search felt broken—users were abandoning apparel queries at a rate two and a half times higher than queries in other categories. The team had already spent significant annotation budget on random sampling over the previous two quarters, with no material improvement in the apparel slice.

The product manager, working with the ML engineer, ran a failure analysis on three months of apparel search logs. They identified three specific failure patterns: queries containing fabric descriptors (such as "linen" or "breathable"), queries using fit language ("relaxed fit," "oversized"), and multi-attribute queries combining color and occasion ("navy formal"). Together, these three patterns accounted for 68% of all apparel query abandonment events but represented only 4% of total search traffic—which is why random sampling had never adequately covered them.

The team ran a two-week targeted labeling sprint: eight hundred examples drawn exclusively from the three failure patterns, labeled by a panel of annotators calibrated specifically on apparel taxonomy. The sprint cost approximately $12,000 in annotation fees and ML engineer time. After retraining on the augmented dataset, the model's apparel-slice MRR improved from 0.61 to 0.79—a gain that matched the overall model benchmark for the first time. Apparel query abandonment dropped by thirty-one percent over the following four weeks.

The lesson was not that targeted labeling is always superior to random sampling—it is that the decision about where to label should be driven by failure analysis, not by convenience. The product manager's contribution was to insist on failure analysis before approving the labeling budget and to frame the sprint as a hypothesis test: we believe these three patterns cause most of our apparel failures; let us spend two weeks testing that hypothesis. That framing kept the sprint focused and made the outcome easy to evaluate.

5.10 Manager's Checklist

- Pipeline: a written handoff contract exists for every zone boundary, and you can point to it
- Quality gates: schema, completeness, and distribution checks are live and have named owners for each alert type
- Alerts: alert routing has been reviewed in the last 90 days to prevent fatigue and ensure coverage
- Feature store: every feature consumed by a production model is registered, described, and has a documented owner
- Training-serving skew: your ML team can demonstrate point-in-time correctness and training-serving parity

- Annotation guide: it is versioned, current, and in the same repository as your training code
- IAA: inter-annotator agreement is reported at least monthly; you know your Kappa score
- Labeling budget: each annotation investment has a documented ROI estimate based on expected model improvement
- Sampling strategy: your current labeling queue is not purely random; you have reviewed the failure analysis
- Slice coverage: you know which input slices are underrepresented in your training data and have a plan to address them
- Data readiness checklist: You run it before every major labeling sprint, model development sprint, or production deployment

5.11 Takeaway

Data strategy is not the infrastructure team's problem—it is the product manager's responsibility. The opening scenario of this chapter—a recommendation model degrading silently for three weeks because no one owned the pipeline contract or the quality gate—is not an engineering failure. It is a governance failure, and governance is a product management function.

The pipeline architecture, quality gates, labeling workflows, feature store discipline, and sampling strategies described in this chapter form a coherent system. Each component supports the others: quality gates protect the feature store; the feature store prevents training-serving skew; annotation governance produces the ground truth that makes quality gates meaningful; sampling strategy ensures that annotation budget closes the gaps that quality gates surface. Managed as a system, these practices transform data from a source of silent risk into a source of durable competitive advantage.

The manager who can walk into a quarterly business review and explain exactly why model performance improved—pointing to a specific labeling sprint, quality gate, or feature definition—is demonstrating operational clarity. That clarity is the deliverable. Start with the data readiness checklist, run it before every significant investment, and build your governance practices one component at a time. The twenty-two-day silent failure that opened this chapter is avoidable. So is the next one.

6 Evaluation and Release Readiness for Generative Systems

The product team had done everything right — or so it appeared —for six weeks of functional testing. Hundreds of user stories verified. Latency benchmarks met. Accuracy scores that looked solid on a holdout set. The generative AI assistant was cleared for launch, and the rollout went smoothly on day one.

By day three, the support queue was filling with complaints. Users had discovered that the assistant answered policy questions with absolute confidence — and was frequently wrong. Not subtly wrong. Categorically wrong. It cited a maternity leave policy revised two years earlier. It quoted a benefits cap that did not exist. One user printed the assistant's output and brought it to HR as proof of entitlement. The team had tested whether the assistant answered questions. They had never tested whether it answered them truthfully, or what would happen when it did not know the answer.

There was no hallucination testing protocol. No adversarial prompt review. No red-team exercise. No rollback criteria tied to fabrication rates. The evaluation framework had been borrowed wholesale from a classical software project. For a deterministic system, that would have been adequate. For a generative one, it was a serious gap — and it cost the team months of remediation work, a damaged reputation with HR and legal stakeholders, and a mandatory re-evaluation before the assistant was allowed back into production.

6.1 Why It Matters

Generative AI systems require a fundamentally different approach to evaluation. Standard software QA is necessary — but it is not sufficient. A traditional QA process verifies that software behaves deterministically: given input A, the system produces output B. Generative systems do not work that way. Their outputs are probabilistic, context-sensitive, and variable. Two identical prompts can produce meaningfully different responses. A system that passes every functional test can still fail catastrophically when exposed to unanticipated real-world inputs.

For a product manager, the implications are direct. Your release-readiness checklist must expand. Your definition of "working" must include not just functional correctness but factual reliability, safety under adversarial conditions, graceful fallback behavior, and measurable hallucination rates. These are not engineering concerns delegated to the model team. They are product decisions that affect user trust, legal exposure, and business outcomes. You own the release gate.

This chapter equips you to build and operate a rigorous, practical evaluation framework for generative AI systems. It covers the key evaluation dimensions — safety, hallucination, adversarial robustness, and UX fallback behavior — along with structured approaches to red-teaming, staged rollout, and rollback criteria. The goal is operational clarity: you should finish this chapter knowing exactly what to test, who tests it, when, and what thresholds trigger a hold.

6.2 The Evaluation Framework for Generative AI

Traditional software evaluation rests on a foundation of determinism: a defined input produces a defined output, and a test either passes or fails. Performance, regression, and integration tests all share this logic. They are effective because the behavior space is bounded and predictable.

Generative AI expands the behavior space enormously. The same prompt, sent to the same model on the same day, may produce different outputs. The model has no memory of prior interactions unless you explicitly provide it. It can be influenced by framing, context length, prior conversation history, and subtle variations in wording. It can produce outputs that are fluent and confident, even when factually wrong. Standard testing does not surface these failure modes.

6.2.1 Why Traditional Metrics Fall Short

Accuracy metrics computed on a holdout dataset are necessary starting points but misleading in isolation. A model may achieve 95% accuracy on a curated benchmark while hallucinating on 12% of real-world queries. Latency benchmarks tell you nothing about output quality. Pass/fail regression tests catch regressions in known behaviors but cannot surface novel failure modes that emerge in production.

For generative systems, the relevant evaluation dimensions include: factual accuracy and hallucination rate; adherence to safety guidelines; resistance to adversarial manipulation; consistency across semantically equivalent prompts; tone and policy compliance; appropriate handling of out-of-scope queries; and UX fallback behavior when the model is uncertain. Each dimension requires its own measurement methodology and its own threshold for release approval.

6.2.2 Building a Multi-Dimensional Evaluation Matrix

A practical evaluation matrix for a generative system maps each evaluation dimension to a measurement method, a responsible owner, a minimum acceptable threshold, and a release hold condition. This matrix is not a one-time artifact. It is a living document that evolves as the system matures and as you discover new failure modes.

Each dimension should be evaluated by an appropriate method: automated scoring for factual accuracy and hallucination rate; structured human review for tone, policy adherence, and safety; dedicated red-team exercises for adversarial robustness; and UX research for fallback behavior and user trust. No single method covers all dimensions. A rigorous evaluation program combines all of them and tracks results over time.

Diagram 6.1 – Evaluation Dimension Matrix for Generative AI

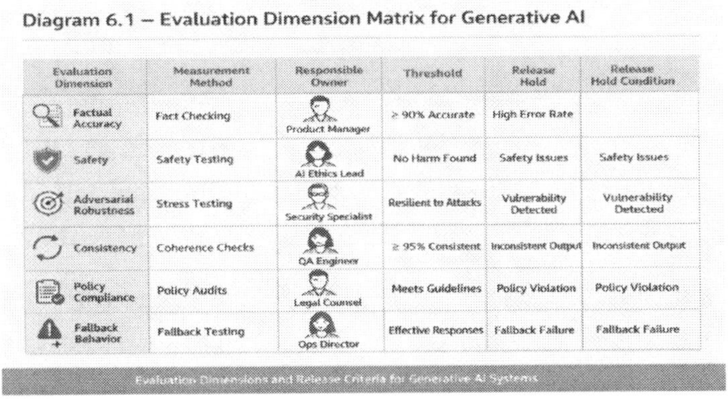

Diagram 6.1 – Evaluation Dimension Matrix for Generative AI

Evaluation Dimension	Measurement Method	Responsible Owner	Threshold	Release Hold	Release Hold Condition
Factual Accuracy	Fact Checking	Product Manager	≥ 90% Accurate	High Error Rate	
Safety	Safety Testing	AI Ethics Lead	No Harm Found	Safety Issues	Safety Issues
Adversarial Robustness	Stress Testing	Security Specialist	Resilient to Attacks	Vulnerability Detected	Vulnerability Detected
Consistency	Coherence Checks	QA Engineer	≥ 95% Consistent	Inconsistent Output	Inconsistent Output
Policy Compliance	Policy Audits	Legal Counsel	Meets Guidelines	Policy Violation	Policy Violation
Fallback Behavior	Fallback Testing	Ops Director	Effective Responses	Fallback Failure	Fallback Failure

Evaluation Dimensions and Release Criteria for Generative AI Systems

6.3 Safety Testing and Adversarial Prompt Testing

Safety testing for generative AI is the systematic effort to discover outputs that could harm users, expose the organization to legal liability, or damage its reputation. It goes beyond functional correctness to ask a harder question: what does this system do when users push it in directions it was not designed for?

Adversarial prompt testing is a structured technique within safety testing. It involves deliberately crafting inputs designed to elicit harmful, false, biased, or policy-violating outputs. Thc goal is not to find edge cases for their own sake, but to understand the system's failure envelope before real users discover it.

6.3.1 Categories of Adversarial Inputs

Adversarial inputs fall into several categories that a product manager should know by name. Prompt injection attempts to override system instructions by embedding directives in user input—for example, by asking the assistant to "ignore all prior instructions." Jailbreaking uses roleplay, fictional framing, or indirect requests to circumvent safety filters. Data exfiltration probes attempt to elicit confidential information present in the context or training data. Social engineering probes test whether the model can be manipulated into producing harmful recommendations.

Bias and fairness probes test for differential behavior across demographic groups, topics, or sensitive categories. Consistency probes submit semantically equivalent questions in different forms to verify that the model does not give contradictory answers. Out-of-scope probes test how the system handles questions it should decline — ensuring that refusal behavior is graceful and informative rather than confusing or evasive.

6.3.2 Structuring Your Safety Test Suite

A structured safety test suite is organized by category and severity. Each test case includes the input, the expected behavior (not necessarily a specific output, but a class of acceptable outcomes), the actual output, a pass/fail judgment, and a severity rating. Severity ratings distinguish between critical failures (outputs that are harmful, discriminatory, or legally problematic), major failures (outputs that are misleading or policy-violating), and minor failures (outputs that are suboptimal but not harmful).

The minimum viable safety test suite for a customer-facing generative system should cover: at least 50 prompt injection attempts across a range of styles; 30 jailbreak scenarios using the most common techniques; 20 data exfiltration probes relevant to the system's knowledge base;

30 bias and fairness probes across the relevant demographic and topical categories; and 20 out-of-scope handling tests. Larger, higher-risk systems require proportionally larger test suites. These are floors, not ceilings.

Diagram 6.2 – Safety Test Suite Structure and Severity Tiers

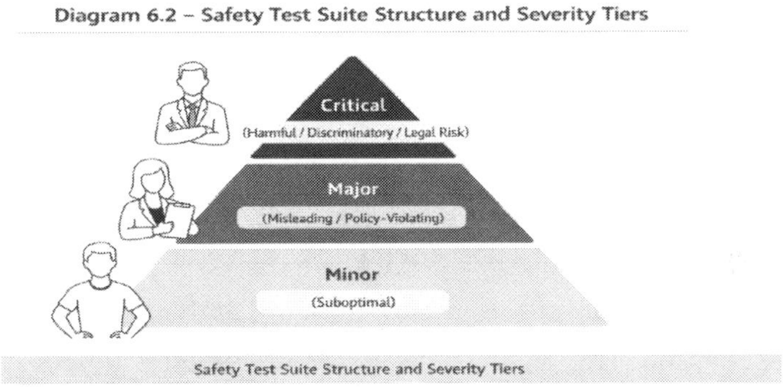

6.4 Operationalizing Red-Teaming in a PM Cadence

Red-teaming is the practice of systematically attempting to break your system before it reaches users. The term comes from military simulation: a "red team" plays the adversary to stress-test plans and defenses. In AI product development, red-teaming means assembling people whose job is to find the worst things your system can do — and documenting what they find.

Red-teaming is not a one-time pre-launch activity. It is an ongoing practice that should be embedded in your regular product cadence. The threat landscape evolves. New use patterns emerge in production. Model updates change behavior in subtle ways. A red-team exercise that was comprehensive at launch may be outdated six months later.

6.4.1 Who Participates in Red-Teaming

Effective red teams are diverse. They include people who understand the system architecture and can probe technical vulnerabilities; domain experts who understand the subject matter well enough to test factual reliability; people from affected communities who can surface bias and fairness issues that technical staff might miss; and, where possible, external reviewers who bring no prior assumptions about the system.

For most enterprise AI products, a practical red team composition is: two to three members of the internal product or engineering team; one or two domain subject-matter experts; at least one stakeholder from a user-adjacent function (customer success, legal, compliance, or HR, depending on the product); and, for high-stakes systems, an external AI safety consultant or a structured crowd-sourced exercise. Budget for external participation early — it is harder to justify retroactively after an incident.

6.4.2 Red-Teaming Schedule and Triggers

Red-teaming should be scheduled at three natural points: pre-launch, as a condition of release readiness; post-major-update, whenever a model version, prompt template, or retrieval system is significantly changed; and on a periodic cadence, typically quarterly for active production systems. These are the minimum cadence checkpoints.

Incident reports, user feedback patterns, or changes in the threat landscape should trigger additional exercises.

Each red-team exercise should produce a structured findings report covering: the number of tests conducted by category; the number and severity of failures found; root-cause analysis for critical and major findings; recommended mitigations; and a sign-off status for each finding (open, mitigated, accepted with documented rationale). Product managers own the disposition of findings. Engineering implements mitigations. Legal and compliance review accepted risks. All of this must be documented before a release gate is passed.

6.4.3 What to Do with Findings

Red-team findings fall into three disposition categories. Critical findings are blocking: the system cannot be released until they are mitigated or formally accepted by an authorized stakeholder (typically VP-level or above). Major findings are conditionally blocking: they must be either mitigated before release or accepted with a documented mitigation plan, monitoring hook, and timeline. Minor findings are tracked in the backlog with priority based on frequency and potential impact.

Do not allow red-team findings to disappear into a spreadsheet. Integrate them into your standard product backlog. Track them through your normal sprint and review cycles. If a finding is accepted rather than mitigated, document the business rationale, the responsible executive, and the monitoring plan. "We know about this risk and have accepted it for these reasons" is a defensible position. "We found this issue and lost track of it" is not.

6.5 Hallucination Metrics and Measurement

Hallucination is the tendency of generative models to produce outputs that are fluent, confident, and factually incorrect. The term is somewhat imprecise — it encompasses a range of failure modes including outright fabrication, plausible-sounding but unsupported claims, incorrect citations, and subtle misstatements. For a product manager, the practical question is simpler: how often does this system tell users something wrong?

Hallucination rate matters because users cannot reliably detect it. Unlike a system error that announces itself, a hallucinated response looks exactly like a correct one. It is grammatically sound, contextually appropriate, and delivered with the same confidence as accurate information. Users who trust the system — especially power users who have found it reliable in the past — are the most vulnerable. They are the least likely to verify its outputs.

6.5.1 Defining Hallucination for Your System

Before you can measure hallucination, you need a working definition specific to your system. For a RAG-based system retrieving from a document corpus, hallucination typically means generating a claim that the retrieved documents cannot support. For a fine-tuned model answering domain-specific questions, it means generating a claim that contradicts the training knowledge base. For a general-purpose assistant, it means generating a claim that is factually incorrect according to authoritative sources.

Your definition should be operationalizable: a reviewer should be able to look at a response and a reference source and make a consistent judgment about whether a claim is hallucinated. Vague definitions produce inconsistent ratings. Inconsistent ratings produce unreliable metrics. Spend time upfront aligning your review team on examples that illustrate the definition — especially borderline cases.

6.5.2 Measurement Methods

Three measurement methods are in common use. Automated fact-checking uses an LLM or specialized model to evaluate whether a reference source supports each claim in a response. It is fast and scalable, but imperfect — the evaluation model can hallucinate itself. Human review is the gold standard: trained reviewers assess a sample of responses against reference documents and score each claim as supported, unsupported, or contradicted. It is accurate but expensive and slow. Hybrid evaluation uses automated scoring to identify candidate hallucinations for human review, combining scale with accuracy.

For release readiness, human review of a representative sample is non-negotiable. Automated methods can supplement ongoing monitoring in production, but you should not gate a release on automated hallucination scores alone. The sample size for human review should be large enough to detect a meaningful hallucination rate with statistical confidence—typically 300 to 500 responses for a system with moderate query diversity, and more for systems with broad topic coverage.

6.5.3 Setting Hallucination Thresholds

Hallucination thresholds depend on risk context. A low-stakes internal productivity assistant — where users are expected to verify outputs and the cost of an error is a minor inconvenience — might accept a hallucination rate of 5% or below. A customer-facing assistant answering questions about benefits, policy, or product features requires a much lower threshold, typically 1% to 2% or below, with near-zero tolerance for hallucinations in high-consequence domains (medical, legal, financial).

Set your thresholds before you measure, not after. Setting thresholds after measurement creates an incentive to rationalize observed rates rather than honestly assess whether they are acceptable. Document the threshold, the rationale, and the risk context. If the system fails to meet the threshold in pre-release evaluation, the release is held until mitigations are in place and the threshold is met.

Diagram 6.3 – Hallucination Measurement Pipeline

Diagram 6.3 – Hallucination Measurement Pipeline

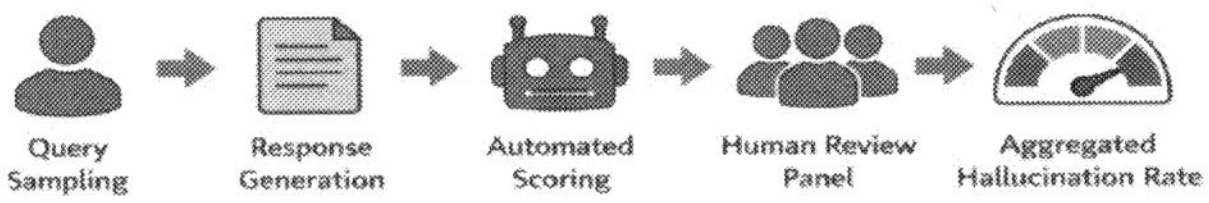

Evaluation process for hallucination measurement

6.6 The Release-Readiness Checklist for LLMs

A release-readiness checklist for a large language model-based product organizes all required validations into a single accountable document. Every item on the checklist must be signed off on before the system advances to the next rollout stage. This is not a courtesy document — it is an operational gate. An incomplete checklist blocks the release.

6.6.1 Functional Tests

Functional tests verify that the system performs its intended tasks at acceptable quality levels. These are the closest generative AI equivalents to traditional software QA.

- Accuracy on benchmark query set meets or exceeds threshold (defined per requirements document)
- Response latency at P50, P95, and P99 meets defined SLA targets
- System handles peak load without degradation (load test results documented)

- Retrieval system (if applicable) returns relevant documents for the representative query set
- Prompt templates produce expected output formats consistently
- Fallback responses trigger appropriately for out-of-scope queries
- Multi-turn conversation context is maintained correctly across the defined session length
- API error handling produces graceful user-facing messages

6.6.2 Safety Tests

Safety tests verify that the system behaves safely under adversarial and edge-case conditions.

- Prompt injection test suite: zero critical failures, major findings resolved or accepted
- Jailbreak test suite: zero critical failures, major findings resolved or accepted
- Bias and fairness evaluation: no statistically significant differential performance across protected categories
- Data exfiltration probes: system does not reproduce confidential training data or system prompts
- Out-of-scope handling: appropriate refusal behavior documented and tested
- Red-team exercise conducted; findings report reviewed and dispositioned
- Legal and compliance review completed; sign-off documented

6.6.3 **Performance Tests**

Performance tests address non-functional requirements such as reliability, cost, and scalability.

- Hallucination rate below defined threshold (human-reviewed sample, methodology documented)
- Consistency score across semantically equivalent prompts meets the defined threshold
- Cost per inference within budget at projected query volume
- System availability target verified under sustained load
- Graceful degradation behavior confirmed when upstream dependencies are unavailable

6.6.4 **UX Fallbacks**

UX fallback tests verify that user experience degrades gracefully when the model is uncertain, incorrect, or unavailable.

- Low-confidence responses trigger appropriate hedging language or escalation prompts
- Error states present actionable guidance, not raw error messages
- Human escalation path is surfaced for high-stakes queries (defined per product spec)
- User-facing disclaimers and limitations are accurate and prominently displayed
- UX research confirms users understand system limitations and appropriate use cases

6.6.5 Monitoring Hooks

Monitoring hooks ensure that production telemetry is in place before launch.

- Hallucination rate monitoring is live, and alerting thresholds are configured
- Safety incident logging captures flagged outputs with user context
- Latency and error rate dashboards are operational
- User feedback mechanism is instrumented and routed to the product team
- Query volume and cost dashboards are operational
- Drift detection for input distribution is configured
- Incident runbook is documented and distributed to the on-call team

6.7 Staged Rollout Plan: Shadow, Canary, Progressive

Even a system that passes every pre-release evaluation check should not go to full production traffic on day one. Staged rollout is the practice of exposing the system to progressively larger user populations, with decision checkpoints at each stage. It gives you an early signal from real user behavior before you are fully committed.

6.7.1 Shadow Mode

Shadow mode is the first stage of the rollout. The generative system processes live production queries in parallel with the existing system, but its outputs are not shown to users. Instead, they are logged, compared against the production system's outputs, and reviewed by the product team. Shadow mode gives you a real-world distribution of queries to evaluate against without exposing users to any risk.

Shadow mode duration depends on query volume. The goal is to accumulate a representative sample—typically 1,000 to 5,000 queries spanning the full range of expected topics and user types. For low-volume systems, this may require several weeks. For high-volume systems, a few days may suffice. Review the shadow outputs against your evaluation dimensions before advancing to Canary.

Decision threshold to advance from shadow: hallucination rate below release threshold on shadow sample; no critical safety failures in shadow outputs; latency within SLA at observed query distribution; cost within budget projection. If any threshold is not met, shadow mode continues until mitigations are applied and the threshold is met.

6.7.2 Canary Release

Canary release exposes the generative system to a small fraction of live users — typically 1% to 5% of traffic, depending on risk tolerance and user volume. Users in the Canary cohort interact with the new system in real time. Their sessions are closely monitored for safety incidents, user-satisfaction signals, and operational metrics.

Canary cohort selection matters. Do not default to random sampling if your user base has meaningful segments. Consider selecting a canary cohort that is representative of your highest-risk user segments — power users who explore edge cases, users in high-stakes domains, and users with accessibility needs. If something goes wrong, you want to know early, with the smallest possible blast radius.

Decision threshold to advance from canary: satisfaction score (CSAT or equivalent) at or above baseline; safety incident rate below defined threshold; hallucination rate in live sessions consistent with pre-release evaluation; no critical escalations from support team. Canary duration is typically one to two weeks for enterprise systems, longer for high-stakes deployments.

6.7.3 **Progressive Rollout**

Progressive rollout expands traffic in increments — 5%, 10%, 25%, 50%, 100% — with a mandatory evaluation window at each stage. Each increment requires that the metrics from the prior stage continue to hold. A single stage that produces anomalous metrics stops the rollout and triggers an investigation. The increment schedule and threshold criteria should be documented before rollout begins, not determined on the fly.

Progressive rollout is not just a risk-management tool. It is a learning opportunity. Real users at scale interact with systems in ways that internal testers do not. Each increment reveals new query patterns, new failure modes, and new UX issues that can be addressed before full production. Budget time for analysis and iteration during progressive rollout — this is not idle time between stages.

Diagram 6.4 – Staged Rollout Flow: Shadow to Canary to Progressive

Diagram 6.4 — Staged Rollout Flow: Shadow to Canary to Progressive

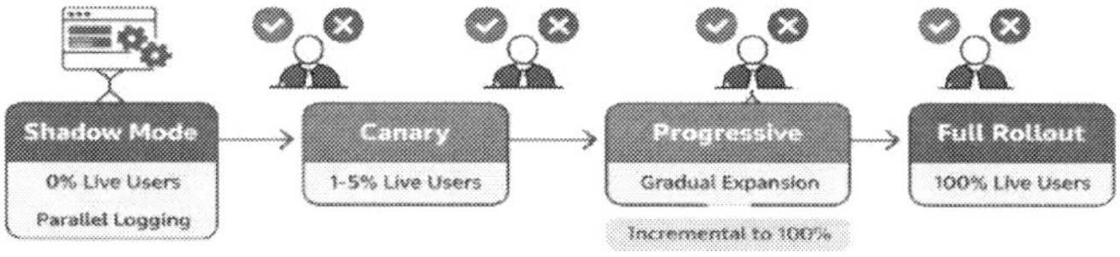

6.8 Rollback Criteria and Decision Points

Rollback criteria define the conditions under which a staged rollout stops and the system reverts to the prior state. They are not optional. A staged rollout without documented rollback criteria is not a controlled rollout — it is an uncontrolled experiment. Rollback criteria must be defined before the rollout begins, not after a problem is detected.

Rollback criteria should be specific and measurable. "The system is performing poorly" is not a criterion for rollback. "The hallucination rate in live sessions exceeds 3% over a 24-hour window" is a rollback criterion. "User satisfaction drops more than 8 percentage points below baseline in the canary cohort" is a rollback criterion. Specificity enables fast, unambiguous decision-making when problems arise.

6.8.1 Automatic vs. Manual Rollback

Some rollback triggers should be automatic: conditions so severe or so clearly defined that human judgment is neither required nor desirable in the moment. If a safety monitoring system detects output matching a critical safety-violation pattern, traffic should automatically route away from the new system, without waiting for a human decision. Configure these automatic triggers before launch and test them in staging.

Other rollback triggers require human judgment: anomalies that need investigation before a decision is made, metrics that are degraded but not yet at threshold, or signals that may reflect a data issue rather than a model issue. For these, define who has rollback authority — typically the product manager or on-call lead — and ensure they are reachable and briefed on rollback criteria before the rollout begins.

6.8.2 Post-Rollback Process

A rollback is not a failure. It is a controlled response to a signal that the system is not yet ready. What matters is what happens next. After every rollback, the team should conduct a structured post-mortem: what metric triggered the rollback? What was the root cause? Was the pre-release evaluation adequate, or did it miss something? What changes are required before the next rollout attempt?

Document rollback events with the same rigor as production incidents. They are leading indicators of evaluation gaps and operational risks. A team that rolls back twice from the same failure mode has a process problem, not a model problem. Use rollback events to drive improvements in your evaluation framework, your release-readiness checklist, and your monitoring configuration.

6.9 Vignette: Red-Teaming Catches a Critical Safety Gap

A financial services firm was preparing to launch an AI assistant for retail investors. The product had passed all functional tests, met latency targets, and scored well on the benchmark accuracy suite. The PM scheduled the red-team exercise as the final pre-launch check, three weeks before the planned release date.

The red team included two internal product engineers, a compliance officer with fifteen years of experience in financial regulation, and an external AI safety consultant. In the first session, the compliance officer tried a prompt that the engineering team had not anticipated: a question framed as a request for tax advice on a specific transaction, combined with a detail that created a regulatory edge case. The assistant responded with a confident, detailed, and technically incorrect analysis that would have constituted unlicensed investment advice under applicable regulations.

Further probing revealed that the system consistently failed on a category of regulatory edge cases where the correct response was to decline and direct the user to a licensed advisor. The failure was not random — it was systematic. The prompt templates did not include adequate guardrails for this category, and the benchmark suite had not included regulatory edge cases.

The PM held the release. The team spent two weeks revising the prompt templates, adding regulatory guardrails, expanding the benchmark suite with 80 new edge-case queries, and running a second red-team exercise to verify the mitigations. The second exercise found no critical failures. The release was delayed by three weeks.

The alternative — proceeding without the red-team exercise, or discovering the failure in production — would have exposed the firm to regulatory action, potential fines, and a mandatory product recall. The three-week delay was not a cost. It was insurance. Thorough pre-release evaluation is not an obstacle to velocity. It is the mechanism by which velocity is made safe.

Diagram 6.5 – Pre-Release Evaluation Gate: Flow from Red-Team to Release Decision

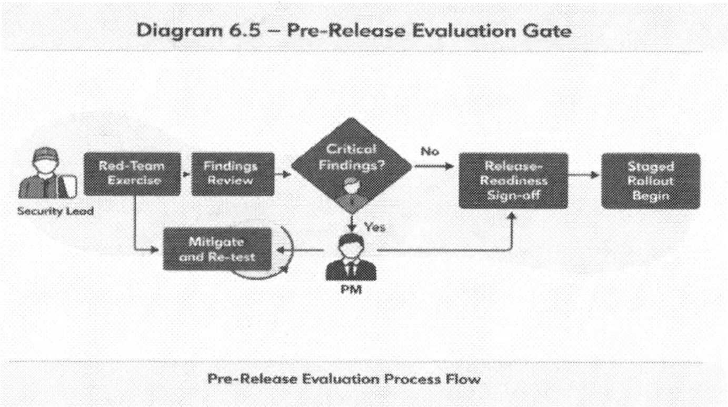

6.10 Manager's Checklist

Use this checklist before each release gate decision for a generative AI system.

- **Evaluation framework defined:** All evaluation dimensions identified, methods assigned, thresholds documented.
- **Functional tests complete:** Accuracy, latency, load, retrieval, and prompt consistency all verified.

- **Safety test suite executed:** Prompt injection, jailbreak, bias, data exfiltration, out-of-scope handling all tested; findings dispositioned.
- **Red-team exercise conducted:** Diverse team, structured findings report, all critical findings resolved or formally accepted.
- **Hallucination rate measured:** Human-reviewed sample, defined methodology, result below threshold; threshold documented before measurement.
- **Release-readiness checklist complete:** Functional, safety, performance, UX fallback, and monitoring hook sections all signed off.
- **Staged rollout plan documented:** Shadow mode criteria, canary cohort definition, progression thresholds, and rollback criteria all specified.
- **Rollback criteria defined:** Specific, measurable thresholds documented; automatic triggers configured; rollback authority assigned.
- **Monitoring live before launch:** Hallucination rate, safety incidents, latency, cost, and user satisfaction dashboards are all operational.
- **Legal and compliance sign-off obtained:** Review documented, accepted risks formally recorded, and responsible executives notified.
- **Post-rollback process defined:** Structured post-mortem template ready; rollback events treated as evaluation improvement inputs.

6.11 Takeaway

Generative AI systems fail in ways that classical software does not. They fabricate with confidence. They can be manipulated into violating their own safety guidelines. They behave differently at scale than they did in testing. A product manager who brings a traditional software QA mindset to a generative AI release is taking a risk they may not understand until it materializes.

The evaluation framework described in this chapter is the operational clarity that responsible generative AI product ownership requires. Multi-dimensional evaluation matrices, structured safety test suites, red-team exercises embedded in the product cadence, human-reviewed hallucination measurement, staged rollout with documented thresholds, and defined rollback criteria — none of these is optional. Each one closes a gap that, left open, produces incidents.

The product managers who build durable reputations in generative AI will be the ones who treat evaluation as a core product competency, not a pre-launch checklist to rush through. Their systems will fail less often. When they do fail, the failures will be smaller and faster to remediate. That is not a coincidence. It is the result of operational discipline applied before the release gate, not after.

7 Human-in-the-Loop: Patterns that Scale

It is 9:00 a.m. on a Monday, and the inbox of Meridian Insurance's claims-processing team already holds 1,400 AI-generated recommendations awaiting human sign-off. Three reviewers are assigned to the queue. By the time the last claim clears on Friday afternoon, the AI system that was supposed to cut processing time by forty percent has instead created a new bottleneck—a waiting room staffed by exhausted humans whose only job is to rubber-stamp outputs they barely have time to read. The data scientists call it a success because the model's accuracy is ninety-three percent. The operations manager calls it a disaster because throughput has not budged.

The root cause is not the model. It is the oversight architecture. The team designed a human-in-the-loop (HITL) review as a simple binary gate: every AI output must be approved before it reaches a customer. That design made sense during the pilot, when volume was low, and trust needed to be established. But as usage scaled from 200 claims per week to 1,400, the binary gate became a wall. Nobody had asked the harder question: at what point does human review add enough value to justify its cost, and how do we route the easy cases differently from the hard ones?

That question is what this chapter is about. HITL is not a checkbox on a compliance form. It is a design challenge that sits at the intersection of workflow engineering, risk management, and organizational behavior. Managers who treat HITL as an afterthought will build bottlenecks. Managers who treat it as a first-class product problem will build systems that are both safe and scalable.

7.1 Why It Matters

Human oversight of AI systems is not optional in regulated industries or anywhere that decisions affect people's lives, livelihoods, or rights. But the form that oversight takes—who reviews what, when, and how—determines whether your AI investment delivers operational value or redistributes the manual work into a new queue. The Meridian scenario is not unusual. It is the default outcome when HITL is designed reactively rather than proactively.

The core insight that shapes everything in this chapter is that HITL is not binary. There is a spectrum between "humans review everything" and "the model acts autonomously," and the right position on that spectrum varies by output type, confidence level, risk profile, and operational capacity. A manager's job is to map that spectrum explicitly, set defensible rules for where each case falls, and then build the metrics and review loops that allow the rules to improve over time.

This chapter provides the pattern catalog, the threshold-setting methodology, the workflow UX principles, the metrics framework, and the staffing model you need to design a HITL that scales. It closes with a vignette showing how a product manager used tiered HITL to maintain safety while growing usage by an order of magnitude, and a Manager's Checklist you can adapt for your own program.

7.2 The HITL Pattern Catalog

Every HITL design is a combination of patterns. Understanding the catalog lets you mix and match deliberately rather than defaulting to whichever pattern your pilot used. There are four foundational patterns, and most production systems deploy two or three of them simultaneously.

Diagram 7.1 – The HITL Pattern Catalog

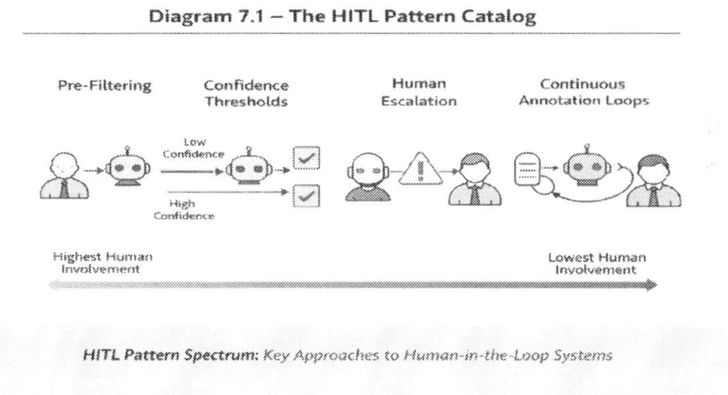

7.2.1 Pre-Filtering

Pre-filtering applies rules before the model runs. It identifies cases that should never enter the automated pipeline—because they are too complex, too sensitive, or too novel for the current model. A claims system might pre-filter any claim above a certain dollar threshold, any claim involving litigation flags, or any claim type the model has seen fewer than fifty times in training.

Pre-filtering keeps high-stakes decisions in human hands without letting them contaminate the model's confidence metrics on cases it does handle well. It also protects the model from inputs that would force overrides and inflate error rates. The design question is not whether to pre-filter but where to draw the line. Start conservative—filter out more than you think necessary—and loosen the boundary as the model proves itself on the edge cases you initially excluded.

Pre-filter criteria should be reviewed quarterly. A common failure mode is building a pre-filter that never changes. As model capability improves, cases that once required human judgment may become automatable, and failing to update the pre-filter means you are paying for human review you no longer need.

7.2.2 Confidence Thresholds

Confidence thresholds are the workhorses of scalable HITL. The model produces an output and a confidence score. Cases above the threshold go straight to action or downstream processing. Cases below the threshold are sent for human review. The threshold is a policy lever, not a fixed truth, and setting it correctly is one of the most consequential decisions an AI product manager makes.

High thresholds mean fewer cases reach reviewers, which reduces costs and turnaround time, but also accepts more model errors in borderline cases. Low thresholds mean more cases go to review, which increases safety but also increases workload and can recreate the bottleneck problem. The right threshold depends on the cost of a false positive versus a false negative in your specific context. This question is ultimately a business decision, not a technical one.

Two-tier thresholds work well in practice. Cases above the upper threshold auto-approve. Cases below the lower threshold are auto-rejected or escalated. Cases in the middle band are sent for human review. This design focuses reviewer attention on genuinely ambiguous cases, which is both more efficient and more intellectually engaging for the reviewers.

7.2.3 Human Escalation

Human escalation routes specific case types to human review based on business rules rather than model confidence. These rules can be domain-driven (any decision involving a protected class), operationally driven (any case flagged by a complainant in the last thirty days), or compliance-driven (any output subject to regulatory audit). Escalation rules live outside the model and should be governed by the business, not the data science team.

The escalation path should be explicit in your workflow documentation. It should specify not just which cases escalate but who receives them, what context they receive, what their decision authority is, and what happens to the model when the escalation decision differs from the model's recommendation. Every escalation is a training signal, and failing to capture it wastes a valuable source of data for model improvement.

7.2.4 Continuous Annotation Loops

Continuous annotation loops treat human review as an ongoing data generation activity rather than a one-time quality gate. Reviewers do not just approve or reject outputs—they provide structured feedback that flows back into the model training pipeline. This pattern is most valuable when the model is still learning, when domain knowledge is evolving rapidly, or when edge cases are frequent enough to require systematic handling.

Annotation loops require more infrastructure than simple escalation. You need a feedback UI, a pipeline to ingest annotations, a process to review annotation quality, and a schedule for retraining. But they compound in value over time: each annotation cycle improves the model, reducing the volume of cases that need review and giving reviewers more bandwidth to focus on the genuinely hard cases, producing higher-quality annotations. This is the flywheel that Meridian Insurance was missing.

Diagram 7.2 – Continuous Annotation Loop Flywheel

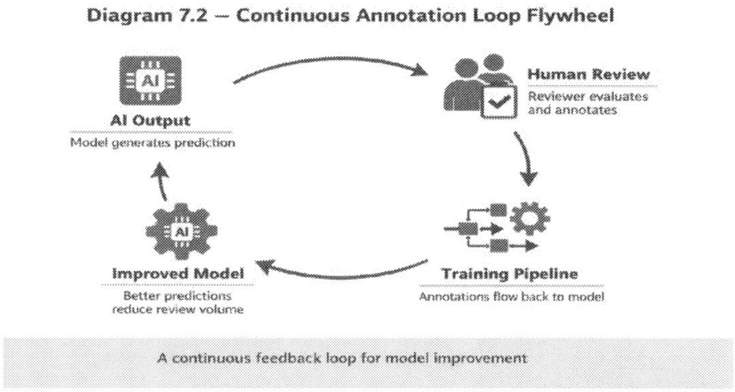

7.3 Confidence Thresholds and Escalation Rules

7.3.1 Setting Thresholds

Threshold-setting begins with a decision matrix that quantifies the cost of two types of error: false positives (the model is confident but wrong) and false negatives (the model is uncertain but would have been right). In a claims context, a false positive might mean approving a fraudulent claim. A false negative might delay a legitimate claim, increase processing costs, and cause customer dissatisfaction. These costs are not symmetric, and the threshold should reflect their actual ratio.

A practical starting approach is to pull a stratified sample of historical outputs and ask reviewers to label them by true outcome. Plot the model's confidence scores against true outcomes and identify the confidence band where errors cluster. Set your lower threshold just above the band's floor. Set your upper threshold at the point where the error rate drops to an operationally acceptable level. These numbers will surprise you—models are often miscalibrated, meaning their stated confidence does not match their actual accuracy.

Calibration is not a one-time exercise. Confidence score distributions shift as input data changes, model versions are updated, and business rules evolve. Build a monthly calibration review into your operating rhythm and treat threshold drift as a risk that must be actively monitored. An auto-approval threshold that was safe at launch may be dangerously permissive six months later if the underlying data distribution has shifted.

Diagram 7.3 – Confidence Threshold Decision Matrix

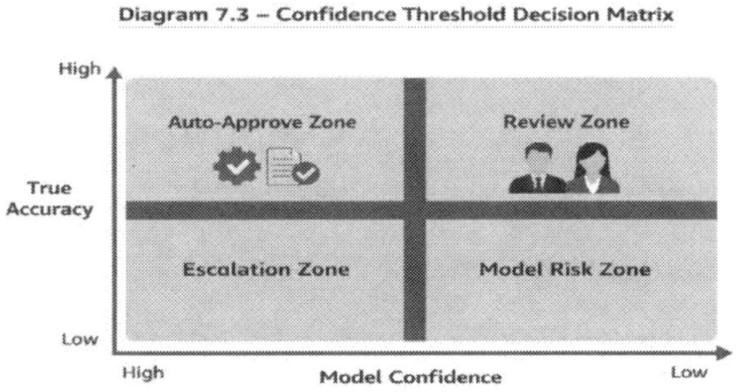

7.3.2 Calibrating Over Time

Calibration improvement follows a predictable arc. In the early months, thresholds will be conservative, and human review volume will be high. As the annotation loop generates more data and the model is retrained, accuracy improves in the zones near the threshold boundaries, and those boundaries can be moved to reduce review load. Track this arc explicitly—plot reviewer volume against model version over time so you have evidence to show stakeholders when arguing for or against staffing changes.

Inter-reviewer agreement is a leading indicator of threshold quality. When reviewers frequently disagree about cases in the middle band, it usually means one of two things: the model is surfacing genuinely ambiguous cases, which is correct behavior, or the threshold band is too wide and capturing cases that should be auto-routed. Distinguishing between these requires reviewing the content of disagreement cases rather than just the count.

Set explicit review triggers for threshold recalibration: a five-point drop in inter-reviewer agreement, a ten-percent increase in escalation rate, or a post-incident review that reveals threshold-related errors. Without explicit triggers, calibration becomes ad hoc and reactive, which is precisely the failure mode you are trying to avoid.

7.4 Reviewer Workflow UX Design

7.4.1 Minimizing Cognitive Load

Reviewer cognitive load is the hidden variable in HITL performance. A fresh, focused reviewer will catch errors that an exhausted reviewer will miss. Yet most HITL workflows are designed with no attention to the cognitive experience of the people doing the reviewing. They present raw model outputs in a table, require reviewers to switch between multiple systems to gather context, and provide no structure for the decision itself. The result is inconsistent quality that masquerades as model performance variance.

The single most effective technique for reducing cognitive load is progressive disclosure. Show the reviewer only what is necessary to make a decision on hand. If the model is reviewing a claim, do not dump the full claim file on the screen—show the model recommendation, the key features that drove it, and the two or three pieces of contextual information a human would actually need to override or confirm. Make additional context available on demand, not by default.

Decision fatigue is real and measurable. Reviewers make systematically worse decisions after prolonged sessions. Design workflow limits include maximum sessions per day, mandatory breaks, and rotation schedules that prevent any single reviewer from becoming the sole point of failure for high-stakes case types. Track reviewer accuracy by time of day and session duration to quantify the degradation curve for your specific workflow.

7.4.2 Task Design

The structure of the review task itself matters enormously. Binary approve/reject decisions are fast but produce low-signal feedback. Structured decisions that ask the reviewer to select a reason code, rate their confidence, and flag cases for follow-up produce higher-signal feedback but take longer. The right balance depends on the downstream use of the feedback. If annotations feed model retraining, invest in a richer structure. If the goal is simply quality control, binary decisions with mandatory escalation flags may be sufficient.

Randomized quality checks—where a percentage of already-approved cases is re-presented to the same or different reviewer—are essential for maintaining annotation quality over time. Without quality checks, reviewer accuracy tends to drift, especially when reviewers believe their decisions are not being monitored. Frame quality checks as calibration support rather than surveillance: reviewers should understand that the checks help the organization learn, not catch individuals making mistakes.

7.4.3 Interface Patterns

Effective reviewer interfaces share a set of design patterns that reduce friction and improve decision quality. First, co-location: all information needed to make the decision should be on a single screen without scrolling or tab-switching. Second, anchoring: show the model recommendation prominently but below the key contextual information, so reviewers form their own assessment before seeing the model's. This reduces anchoring bias, in which reviewers unconsciously defer to the model even when they have information the model lacks.

Third, friction by design: for high-stakes overrides, require the reviewer to type a brief justification rather than clicking a button. This slows the decision slightly but dramatically improves the quality and auditability of overrides. Fourth, queue management: show reviewers how many cases remain in the queue and their expected completion time, but design the display to encourage steady throughput rather than rushing. A real-time counter that drops every second creates anxiety; a projected completion time is calmer and equally informative.

Diagram 7.4 – Reviewer Workflow Interface Layout

Diagram 7.4 — Reviewer Workflow Interface Layout

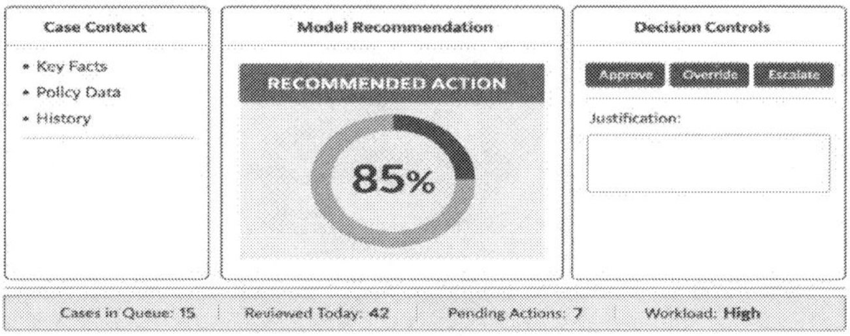

7.5 HITL Metrics That Matter

HITL programs generate a rich set of metrics, and it is easy to measure the wrong ones. Teams often focus on throughput—cases reviewed per hour—because it is easy to collect and easy to present to leadership. But throughput alone tells you nothing about quality, and optimizing for throughput at the expense of quality is how bottleneck problems get replaced by accuracy problems.

7.5.1 Reviewer Throughput

Throughput is necessary but not sufficient. Measure it at the individual reviewer level, the team level, and by case type. Throughput that is significantly higher for one reviewer than for others is a warning sign: either that reviewer is rushing, or the other reviewers are over-deliberating. Neither extreme is healthy. Use throughput variation as a diagnostic, not a KPI you try to maximize.

Queue depth and queue age are more actionable than raw throughput. Queue depth tells you whether review capacity matches incoming volume. Queue age tells you whether specific case types are being deprioritized, which often happens informally when reviewers route around difficult cases. Set service-level agreements by case type and monitor compliance weekly.

7.5.2 Annotation Quality

Annotation quality measures whether the labels that reviewers apply are correct. This requires a ground-truth mechanism: a set of cases with known outcomes that reviewers do not know are test cases. Compare reviewer labels against known outcomes to produce a per-reviewer accuracy score. Aggregate across the team to produce an overall annotation quality index.

Annotation quality typically starts high when reviewers are fresh and the task is novel, then dips as routine sets in, and then can recover if the task design includes variety and feedback. Tracking the quality trend over time is more useful than any point-in-time measurement. A declining trend is an early warning that reviewer calibration or engagement needs attention.

7.5.3 Inter-Reviewer Agreement

Inter-reviewer agreement (IRA) measures how often two reviewers, given the same case, reach the same decision. High IRA indicates clear decision criteria and well-calibrated reviewers. Low IRA indicates ambiguous decision criteria, under-specified escalation rules, or reviewer drift. IRA should be measured monthly using a sample of cases reviewed by multiple reviewers.

Do not target 100 percent IRA. In genuinely ambiguous cases, disagreement is the correct outcome—it signals that the case exceeds the model's reliable operating envelope and should trigger a protocol review, not a performance conversation. The goal is a high IRA on cases that fall clearly within policy, and structured disagreement protocols for cases that do not.

7.5.4 Model Improvement per Annotation

This metric measures the return on your annotation investment: how much does each human label improve model performance? In the early stages of a HITL program, this number is high—the model has much to learn, and every annotation adds signal. Over time, as the model matures, the return per annotation decreases. This curve is the analytical foundation for decisions about when to reduce review volume and when to reallocate annotation effort to new edge-case domains.

Track model improvement per annotation by running before-and-after evaluations on a holdout set after each retraining cycle. Express the result as the change in your primary model metric per thousand annotations. When this number falls below your cost-per-annotation threshold, you are in the regime of diminishing returns, which we address in a later section of this chapter.

Diagram 7.5 – HITL Metrics Dashboard

Diagram 7.5 – HITL Metrics Dashboard

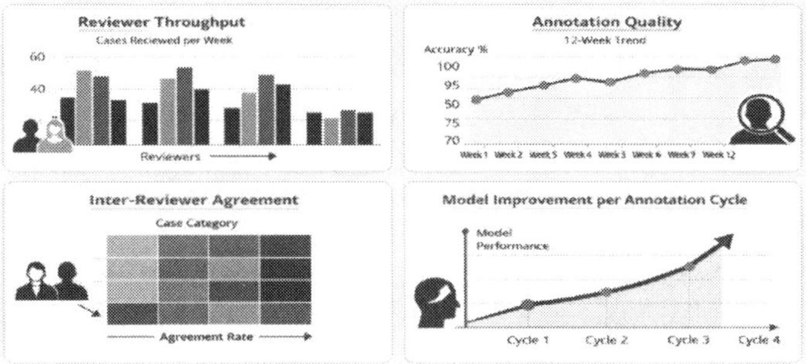

7.6 Costing and Staffing Human Review

HITL is not free, and the organizations that treat it as free are the ones that get caught short when volume scales unexpectedly. Building a defensible cost model for human review is a core product management responsibility, not something to delegate to finance after the fact.

The cost model has four components: labor (reviewer time at fully loaded cost), tooling (the review interface and annotation platform), quality assurance (the calibration infrastructure and QA sampling overhead), and opportunity cost (cases delayed by review that could have been auto-approved without an increase in material risk). Most programs measure only labor and tooling. Ignoring the other two understates both the cost of HITL and the value of investing in threshold improvement.

7.6.1 Staffing Models

There are three staffing models for human review, each with different cost and flexibility profiles. The dedicated model employs full-time reviewers who work exclusively on the HITL queue. This produces the highest reviewer expertise and the most consistent annotation quality, but is expensive and inflexible when volume is unpredictable. The embedded model distributes review responsibilities among subject-matter experts who also handle other work. This leverages existing domain knowledge but creates scheduling complexity and risks review becoming deprioritized when workload spikes.

The hybrid model uses a small core of dedicated reviewers for high-complexity and high-stakes cases, supplemented by embedded reviewers for standard cases during high-volume periods. This is the most common model in mature programs because it balances cost efficiency with quality consistency. The key to making the hybrid model work is clear case routing—the rules that determine which cases go to dedicated reviewers and which go to embedded reviewers must be explicit, documented, and audited regularly.

7.6.2 Budget Planning

Budget HITL labor as a function of review volume, not as a fixed headcount. Project review volume using your confidence threshold settings and historical intake. Apply a reviewer throughput rate that accounts for task complexity and decision fatigue, not just raw speed. Build in a twenty-percent buffer for volume spikes, quality remediation, and calibration overhead. Run the model quarterly and update it when threshold settings change, case mix changes, or intake volume trends diverge from projections by more than 15%.

Track cost per annotation as a unit economics metric. Divide the total HITL operating cost (labor plus tooling plus QA overhead) by the total cases reviewed in the period. This number should decrease over time as automation within the HITL workflow improves and as model improvement reduces the volume of cases requiring review. If cost per annotation is increasing, something structural has changed—case complexity is rising, reviewer efficiency is declining, or tooling overhead has grown—and you need to diagnose which.

7.7 Automation Opportunities Within HITL

HITL is not a binary choice between full automation and full human review. Within the review workflow itself, there are often dozens of sub-tasks that can be automated without compromising the quality or defensibility of the human decision. Identifying and automating these sub-tasks is one of the highest-leverage investments an AI product manager can make in operational efficiency.

7.7.1 Information Retrieval and Pre-Population

The single biggest time cost in most review workflows is not the decision itself—it is the time spent gathering the information needed to make the decision. Reviewers switch between systems, manually pull records, and reconstruct context from scratch for each case. Automating this retrieval and pre-populating the review interface with the relevant context can reduce per-case review time by thirty to sixty percent in data-rich environments.

Pre-population is not just a time-saver. It reduces the risk that reviewers will make decisions based on incomplete information because a relevant record was too time-consuming to retrieve. When you pre-populate, be selective: surface the information that drives decisions ninety percent of the time, and make additional context available on demand. Surfacing everything creates information overload and recreates the cognitive load problem you are trying to solve.

7.7.2 Automated Consistency Checks

Many review errors are not judgment errors—they are consistency errors. The reviewer approved a case that is materially identical to three cases that were rejected last week. Automated consistency checks can flag these conflicts before the reviewer submits a decision, prompting review of the prior decisions and reducing both error rate and inter-reviewer disagreement.

Consistency check engines do not need to be sophisticated. A rules-based system that identifies structurally similar cases from recent history and surfaces prior decisions is often sufficient. The goal is not to make the decision for the reviewer but to provide the context that prevents inconsistency. Reviewers who see similar prior cases make better-calibrated decisions and generate more useful annotations.

7.7.3 Routing and Triage Automation

Within the review queue, not all cases are equally urgent or equally complex. Automated routing assigns cases to reviewers based on case type, reviewer expertise, and queue depth, ensuring that complex cases go to experienced reviewers and time-sensitive cases are prioritized appropriately. This is a form of HITL within HITL: automation that improves the efficiency of human review rather than replacing it.

Routing automation should be transparent to reviewers. They should know why a case was assigned to them and what the expected review parameters are. Opaque routing algorithms that assign cases without explanation reduce reviewer trust and make it harder to identify and fix routing errors. Invest in a routing rationale field in your review interface, even if it only displays a simple label such as "assigned due to complexity flag" or "assigned due to domain expertise match."

7.8 Measuring Diminishing Returns on Human Review

One of the most politically difficult conversations in AI product management is when to reduce human oversight. The instinct—especially in regulated industries—is that more human review is always safer. This is not true. At a certain scale of operation, adding more human review introduces its own errors: reviewer fatigue, inconsistency, and the distraction of reviewing cases the model handles well, which reduces attention available for the cases where human judgment actually matters.

Diminishing returns manifest in two ways. The first is the annotation return curve described in the Metrics section: each additional annotation produces a smaller improvement in model performance. The second is reviewer performance degradation: as queue volume grows, reviewers spend more time on routine cases and have less cognitive bandwidth for genuinely difficult ones.

7.8.1 The Diminishing Returns Curve

Build a diminishing returns model by plotting model accuracy against cumulative annotation volume, segmented by case type. For most models, accuracy improves rapidly in the first few thousand annotations, then the curve flattens. The inflection point—where the slope of the improvement curve drops below your marginal cost per annotation—is the threshold for reducing annotation effort in that case category. This threshold varies by case type, and different parts of your HITL program will hit diminishing returns at different times.

Communicating diminishing returns to stakeholders requires framing. Do not say "we are reducing human review." Say "we are redirecting human oversight to where it adds the most value." Show the annotation return curve. Show the projected accuracy at different review volumes. Identify the specific case types in which review will be reduced and the quality metrics to be monitored to confirm that the reduction is safe. This framing turns a politically charged conversation into a data-driven operational decision.

7.8.2 Governance for Review Reduction

Any reduction in human review should be subject to a formal change-control process. Document the current review volume and accuracy baseline. Specify the reduction being proposed (from what to what, for which case types). Identify the monitoring plan that will detect if accuracy degrades following the change. Define the rollback criteria: what would trigger reinstatement of review for this case type? Get sign-off from the relevant business and compliance stakeholders before implementing.

This process sounds bureaucratic, but it is actually protective—for the organization and for the product manager. It creates an audit trail that demonstrates responsible stewardship of the AI system. If a post-reduction error is investigated, the documentation shows that the decision was deliberate, evidence-based, and appropriately governed. The absence of documentation leaves the impression that the review was reduced carelessly, even if it was not.

7.9 Vignette: Tiered HITL at Meridian Insurance

Twelve months after the Monday-morning inbox crisis, Meridian Insurance's claims system looks very different. The product manager, Sasha, worked with the operations team to redesign HITL from scratch using the tiered pattern.

In the first month, Sasha conducted a case audit. She pulled a stratified sample of 500 claims from the previous quarter. She asked three experienced reviewers to label them independently, using a structured decision template that captured not just approve/reject but also a five-category reason code. This produced three things: an inter-reviewer agreement baseline (seventy-eight percent), a confidence calibration dataset (mapping model scores to actual accuracy), and a categorization of the case mix into four types based on complexity and risk.

Armed with the categorization, Sasha defined a three-tier routing scheme. Tier 1—low complexity, high confidence—was designated for auto-approval, accounting for 40% of case volume. Tier 3—high risk or novel case type—was routed to a dedicated senior reviewer team regardless of model confidence. Tier 2—everything in between—went to the standard review queue with a structured decision interface.

The interface redesign was Sasha's highest-leverage intervention. Working with UX, she eliminated system switching that had cost reviewers an average of 4 minutes per case. The new interface pre-populated the three contextual fields that drove eighty-five percent of decisions. It anchored reviewers with the contextual summary before showing the model recommendation. For overrides, it required a two-sentence rationale, which flowed directly into the annotation pipeline.

Within ninety days, review volume had dropped by forty-two percent even as total case volume grew by thirty percent—a combined efficiency improvement of more than sixty percent. Inter-reviewer agreement climbed to eighty-seven percent. Reviewer satisfaction scores, which Sasha had started tracking because she believed reviewer experience was a leading indicator of output quality, improved by twenty-two points. The model itself improved by 4.1 points of accuracy on the Tier 2 case distribution after the first retraining cycle on the new annotations.

Twelve months in, Sasha presented the executive team with a projection of the diminishing-returns curve for Tier 2 cases. She proposed moving an additional 15% of case volume to Tier 1, based on the calibration data, reducing annual HITL labor costs by an estimated $300,000 while maintaining a documented accuracy floor above the regulatory minimum. The proposal was approved. The compliance team, which had initially resisted any reduction in review, signed off because Sasha had prepared the governance documentation and the monitoring plan before seeking approval.

The lesson from Meridian is not that automation is better than human review. It is that thoughtful design of the boundary between automation and human review that creates value neither can produce on its own. Sasha did not reduce human oversight—she made human oversight more effective by concentrating it where it matters most.

7.10 Manager's Checklist

Use this checklist when designing, auditing, or improving a HITL program:

- **Pre-filtering:** Document explicit criteria for cases that bypass the model entirely and route to humans. Review these criteria quarterly as the model's capabilities evolve.
- **Confidence thresholds:** Set upper and lower thresholds using a calibration dataset. Express them in terms of false-positive and false-negative costs, not just accuracy percentages.

- **Two-tier thresholds:** Separate the auto-approve zone (above upper threshold) from the review zone (middle band) and the auto-reject or escalation zone (below lower threshold).
- **Escalation rules:** Define non-model-based routing rules (case type, dollar amount, complainant history, regulatory sensitivity) and document them outside the model, under business governance.
- **Annotation loop:** Ensure every reviewer's decision flows back into a structured annotation pipeline. Verify that overrides include reason codes and that annotations meet your quality standard before reaching the training pipeline.
- **Reviewer interface:** Apply the progressive disclosure, anchoring, friction-by-design, and queue management patterns. Confirm that context retrieval requires no system switching.
- **Cognitive load controls:** Set session length limits, mandatory breaks, and rotation schedules. Track reviewer accuracy by time of day to quantify fatigue effects.
- **HITL metrics:** Track reviewer throughput (by type and individual), annotation quality (against known ground truth), inter-reviewer agreement (monthly sample), and model improvement per annotation (per retraining cycle).
- **Staffing model:** Budget HITL labor as a function of projected review volume, not fixed headcount. Update the model quarterly and when threshold settings change.

- **Automation within HITL:** Identify sub-tasks in the review workflow that can be automated without reducing decision quality: context pre-population, consistency checks, routing, and triage.
- **Diminishing returns monitoring:** Plot model accuracy against cumulative annotation volume by case type. Identify the inflection point and develop a governance process to reduce reviews in mature case categories.
- **Review reduction governance:** Document baseline, proposed change, monitoring plan, and rollback criteria. Obtain compliance and business sign-off before implementing.

7.11 Takeaway

Human-in-the-loop review is one of the most important design problems in AI product management, and one of the most frequently underdesigned. The organizations that treat HITL as a binary gate—either every output is reviewed, or no output is reviewed—will encounter one of two failure modes: a bottleneck that defeats the operational purpose of automation, or an accountability gap that creates risk without anyone noticing until something goes wrong.

The organizations that treat HITL as a design challenge—building a tiered pattern catalog, calibrating thresholds with data, designing reviewer workflows for cognitive sustainability, tracking the right metrics, automating the sub-tasks within review, and governing the transition to higher automation—will build systems that are simultaneously safer and more operationally efficient than either extreme.

The manager's role in HITL is not to choose between human oversight and AI efficiency. It is to design the architecture that makes them complementary. Every decision in this chapter—where to pre-filter, how to set thresholds, how to design the reviewer interface, how to budget staffing, when to reduce oversight—is a management decision with operational, financial, and reputational consequences. Own it as such, and you will build HITL programs that scale with your organization's ambitions rather than against them.

8 Monitoring, Observability, and MLOps for PMs

On a Tuesday morning, the customer-success inbox filled faster than the team could triage it. Users of a retail recommendation engine complained that the suggestions felt random—irrelevant products, outdated seasonal items, and offers that had expired weeks ago. The support manager escalated to the product owner, who escalated to the data science lead, who spent the next four hours tracing the problem back to its source. The underlying model had been drifting silently for three weeks. Purchase patterns had shifted after a major promotional event, but the model had never been retrained. No alert had fired. No dashboard had changed color. The business had been hemorrhaging conversion rate the entire time, and nobody knew.

That scenario is not unusual. It plays out in organizations of every size and every industry, and it almost always has the same root cause: the team built a model, shipped it, and then assumed the work was done. Monitoring was treated as an engineering nicety rather than a product obligation. The product owner had no mechanism for knowing when the model stopped earning its keep.

This chapter equips you with the monitoring mindset, the minimal tooling stack, and the operational playbook to avoid that story. It covers how to connect model health to business outcomes, how to detect drift before users feel it, how to run an incident, and how to decide when it is time to retrain, patch, or retire. You do not need to be an ML engineer to own this layer. You do need to treat it as seriously as you treat your roadmap.

8.1 Why It Matters

AI products degrade silently. A traditional software bug usually surfaces as an error message, a broken workflow, or a clear system failure. A degrading model produces subtly wrong outputs—wrong enough to frustrate users, reduce engagement, or erode trust, but not wrong enough to trigger an obvious error state. The gap between "the model is technically running" and "the model is serving the business" is precisely where monitoring lives.

For the product owner, monitoring is not an infrastructure concern that belongs solely to the MLOps team. It is a product accountability. When the model degrades, the user experience degrades. When the user experience degrades, business outcomes degrade. The chain of causation is direct, and the product owner is the person responsible for the outcome at the end of that chain. If you do not own the monitoring layer, you are flying blind on your most important product dimension.

The business case for investment in monitoring is straightforward. The cost of a missed drift event—lost revenue, customer churn, emergency retrain cycles, reputational damage, incident postmortems—almost always exceeds the cost of the monitoring infrastructure that would have caught it. The question is never whether to monitor. The question is how to monitor efficiently, what to prioritize, and how to build the alerting and response capability to act before users feel the impact.

8.2 Designing the Minimal Monitoring Stack

The phrase "monitoring stack" can make product owners nervous. It conjures images of complex infrastructure, dedicated tooling teams, and six-figure platform licenses. In practice, the minimal viable monitoring stack for most AI products is smaller and simpler than you expect. The goal is not comprehensiveness—it is a signal. You need enough instrumentation to know when something is wrong, what kind of wrong it is, and who should be paged.

Start with four categories of signal. First, business metrics: the outcomes you actually care about—conversion rate, user satisfaction score, task completion rate, revenue per session, or whatever the model is meant to move. Second, model performance metrics: accuracy, precision, recall, F1, AUROC, or the appropriate measure for your model type, calculated on a representative sample of recent predictions. Third, data quality metrics: schema validation, feature distribution statistics, null rates, and upstream pipeline health. Fourth, operational metrics: latency percentiles, throughput, error rates, and infrastructure costs per inference.

Those four categories are not optional. You can simplify within each category, but you cannot drop a category entirely without creating a blind spot that will eventually find you. A team that monitors only business metrics will not know whether model drift, a bad deployment, a data pipeline failure, or seasonality causes a drop in conversion. A team that monitors only model metrics will not know whether a slight decline in accuracy is actually hurting users or is merely statistically insignificant noise. You need all four to triangulate.

Diagram 8.1 – The Four-Layer Monitoring Stack

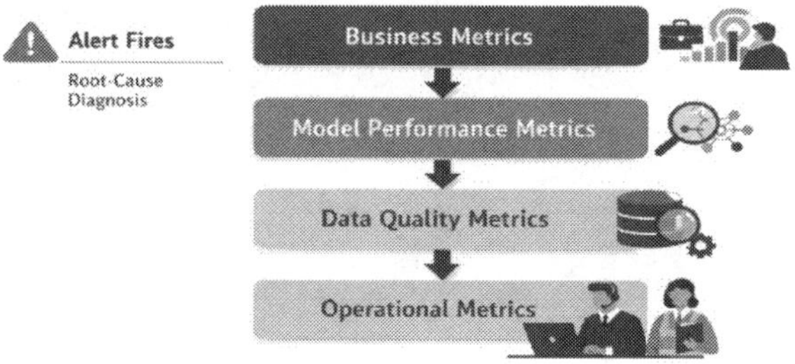

8.2.1 Instrumentation Basics

Instrumentation is the act of making your model and its environment observable. At minimum, this means logging every prediction with its input features, its output, and a timestamp. It means capturing ground truth labels as they become available—whether from user behavior signals, human reviewers, or downstream system events. And it means emitting those logs to a store where you can run queries, compute statistics, and trigger alerts.

The instrumentation work falls under engineering, but the product owner is responsible for specifying what gets logged and which ground-truth signal will be used to measure model quality over time. Ground truth is the hardest part. Some tasks produce near-real-time feedback—a search click signal is available within seconds. Others take days or weeks—a fraud label may not be confirmed until a chargeback is filed. Your monitoring design must account for the latency of your ground truth signal, because that latency determines how quickly you can detect model degradation.

Where ground truth is delayed or sparse, you must rely on proxy metrics—behavioral signals that correlate with model quality without requiring explicit labels. These proxies are imperfect, but they are far better than nothing. A recommendation engine that cannot get explicit relevance labels can use click-through rate, time-on-page, or purchase rate as proxies. A content moderation model can use the appeal rate and escalation volume. Define your proxies in advance, validate their correlation with true quality during your evaluation phase, and monitor them continuously in production.

8.2.2 Monitoring Cadence

Not all monitoring signals need to be checked at the same frequency. A practical cadence tier looks like this: operational metrics (latency, error rate, throughput) are monitored in real time with sub-minute alerting, as they indicate infrastructure failures that users feel immediately. Data quality metrics are checked at pipeline ingestion cadence—typically every batch run or every hour for streaming systems. Model performance metrics and proxy business metrics are evaluated daily for most products; weekly may be acceptable for very low-volume or low-stakes applications. Business outcome metrics are reviewed weekly in operational meetings and monthly in strategy reviews.

Match your alerting cadence to the signal's time-to-impact. A latency spike impacts the next user session. A gradual drift in feature distributions may not impact business outcomes for days or weeks. Alerting on everything at a real-time cadence produces noise that teams learn to ignore. Alerting to the right signals, at the right frequency, produces the situational awareness that makes a fast response possible.

8.3 Business Metrics vs. Model Metrics

One of the most common monitoring failures is tracking model metrics in isolation from business metrics. A team celebrates that their model's accuracy held steady at 94%, while the conversion rate declined by 8 points. A different team panics when accuracy drops from 91% to 89%, even though business outcomes are unchanged. In both cases, the team is looking at the wrong number in isolation.

Model metrics and business metrics are related, but the relationship is not always linear or direct. A small drop in precision may have a negligible business impact if the affected predictions are low-stakes. A large improvement in recall may actually hurt business outcomes if the increase in false positives degrades user experience. You need both layers, and you need a documented hypothesis about how they relate to each other for your specific product and use case.

Diagram 8.2 – Business Metrics vs. Model Metrics Linkage Map

Diagram 8.2 — Business Metrics vs. Model Metrics Linkage Map

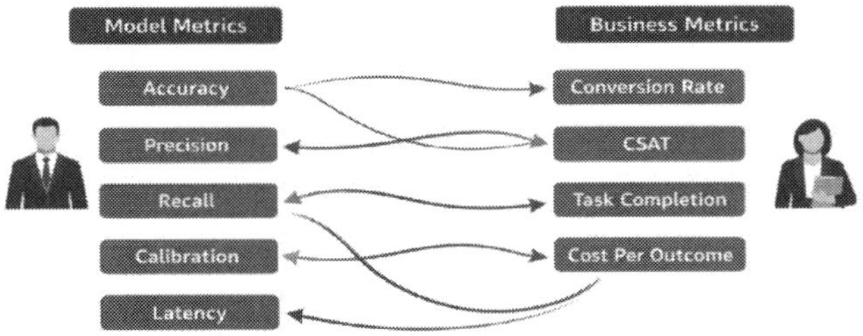

8.3.1 Defining the Connection

Before you launch, you should be able to complete this sentence for your product: "When the model's [metric] drops below [threshold], we expect [business metric] to decline by approximately [range] within [timeframe]." That sentence is your monitoring hypothesis. It is imprecise and will require calibration, but having it forces you to think carefully about the causal chain between model behavior and business outcomes.

Building this connection requires baseline measurement. During your pre-launch evaluation period, deliberately vary model quality in a controlled way—through hold-out tests, degraded model variants, or simulated drift experiments—and measure the corresponding business impact. That data gives you the correlation coefficients you need to set meaningful alert thresholds. Without it, you are guessing, and your guesses will be wrong in ways that generate either alert fatigue or missed incidents.

8.3.2 Calibration as a Business Metric

Calibration deserves special attention because it is often overlooked in standard model evaluations and has a direct business impact. A well-calibrated model means that when it outputs a confidence score of 70%, it is actually correct 70% of the time. Poor calibration—where the model is systematically overconfident or underconfident—affects every downstream decision that uses probability thresholds. If your product uses model confidence to route decisions, prioritize tasks, or gate user access, calibration drift is a business problem, not merely a statistical curiosity. Track it explicitly with calibration error metrics (expected calibration error or reliability diagrams) alongside your standard performance metrics.

8.4 Drift Detection: Data Drift and Concept Drift

Drift is the primary reason AI products degrade over time. The world changes. User behavior changes. The data that arrives at your model changes over time. The relationship between inputs and outcomes changes. Your model, trained on a historical snapshot of reality, becomes progressively less equipped to handle the reality it now encounters. Drift is not a failure of model design—it is an inherent property of deploying a learned system in a dynamic environment. Your job is to detect it early and respond before it becomes a user-visible problem.

There are two distinct types of drift, and they require different detection strategies and different responses.

Understanding the difference is essential for correctly triaging an alert.

8.4.1 Data Drift

Data drift (also called covariate shift) occurs when the statistical distribution of the model's input features changes, even if the relationship between those features and the target variable remains stable. The model was trained on features that looked a certain way, and now the incoming features look different. Examples include a demographic shift in your user base, a change in upstream data-collection methodology, a seasonal pattern that was underrepresented in the training data, or a new marketing campaign that attracts a different customer segment.

Detecting data drift requires baseline statistics. During training or shortly after launch, you capture the distribution of each feature—mean, variance, quantiles, and category frequencies for categorical features. In production, you compare the distribution of incoming features against those baselines at regular intervals. The most common statistical tests for this comparison are the Kolmogorov-Smirnov test for continuous features and chi-squared tests for categorical features. The Population Stability Index (PSI) is also widely used in financial services contexts and provides an intuitive threshold (PSI above 0.2 is conventionally treated as significant drift).

Data drift does not automatically mean your model is performing worse. The model may still generalize well to the new distribution, or the business impact of the drifted features may be small. When data drift is detected, the first response is to investigate whether model performance metrics have also moved. If they have not, the drift may be benign but worth tracking. If performance has degraded, you have a direct causal candidate for retraining.

Diagram 8.3 – Data Drift vs. Concept Drift Detection Flow

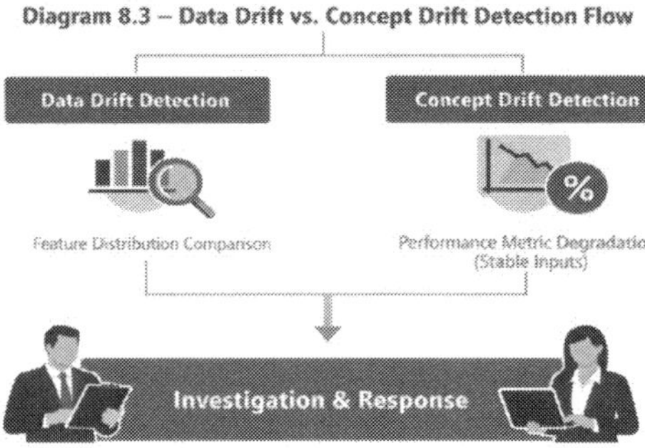

8.4.2 Concept Drift

Concept drift is more insidious. It occurs when the relationship between input features and the target variable changes—when the world changes in a way that makes the patterns the model learned obsolete. The inputs may look fine, the feature distributions may be stable, but the correct answer for a given input is now different from what the model predicts. A fraud detection model trained before a new fraud technique emerges will exhibit concept drift. A demand forecasting model trained before a supply chain disruption will exhibit concept drift. A sentiment classifier trained before a platform's user demographics shifted will exhibit concept drift.

Concept drift can only be detected by comparing predictions against ground truth labels. Without ground truth, you detect concept drift indirectly through proxy signals—behavioral anomalies, spikes in user feedback, or declines in downstream business metrics. This is why investing in ground truth collection infrastructure is so important. The faster you can obtain labels for recent predictions, the faster you can detect concept drift and respond.

Gradual concept drift is particularly dangerous because it moves slowly enough to avoid triggering fixed-threshold alerts while still accumulating into a significant quality problem. Detect it with rolling-window comparisons: compare model performance in the most recent window against a prior reference window, using statistical tests to determine whether the difference is significant. Window sizes should be calibrated to your expected drift velocity—faster-changing domains need shorter windows.

8.4.3 Responding to Drift

When drift is confirmed, your response options span a spectrum of costs and impacts. At the lower end, you can adjust decision thresholds—recalibrate the probability cutoff the model uses to make a binary decision—without retraining. This is fast and cheap, and it can partially compensate for calibration drift. Moving up the spectrum, you can perform targeted retraining on recent data only, which is faster than a full retrain and addresses concept drift without requiring a complete pipeline run. At the higher end, full retraining on an updated dataset, followed by evaluation and a staged rollout, comprehensively addresses both data and concept drift but takes more time and resources. The incident runbook in the following section maps these response options to drift severity classifications.

8.5 Alerting Thresholds and Escalation

Alert design is a product problem, not a platform problem. The question is not what you can monitor—modern MLOps platforms let you alert on almost anything. The question is what you should alert on, at what thresholds, and who responds. Get this wrong in either direction, and you pay a price. Thresholds set too tightly generate noise, training your team to ignore alerts. Thresholds set too loosely allow real problems to fester undetected.

8.5.1 Threshold Setting Principles

The most defensible thresholds are those derived from business impact analysis. If your monitoring hypothesis says that a 5-point drop in precision is associated with a 2% decline in conversion rate, and a 2% conversion decline triggers an urgent business review, then a 5-point drop in precision should trigger an alert. Work backward from acceptable business impact to the corresponding model metric threshold, and you will have thresholds that your stakeholders can understand and endorse.

Where business impact correlation data is unavailable—particularly early in a product's life—use statistical control chart methods. Compute the mean and standard deviation of your performance metrics during a stable baseline period. Alert when a metric falls below the mean minus two or three standard deviations (corresponding to roughly 2.5% and

0.1% false positive rates, respectively). These bounds will need adjustment as the product matures and as you accumulate impact correlation data, but they provide a rational starting point.

Implement at least two alert severity levels. A warning level is triggered at a threshold that indicates potential degradation worth investigating during normal business hours, but not worth a midnight page. A critical level is triggered when a threshold is met, indicating confirmed or imminent user impact and requiring an immediate response, regardless of time of day. The distinction between warning and critical saves your on-call rotation from burnout while ensuring genuine incidents get attention when they matter.

Diagram 8.4 – Alert Severity Matrix and Escalation Ladder

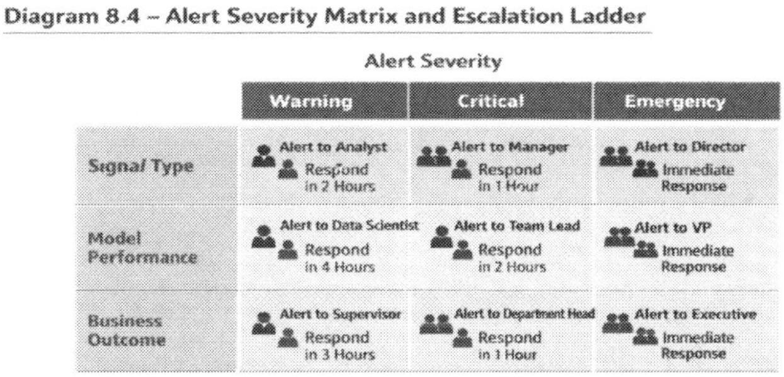

8.5.2 Escalation Design

Escalation design determines what happens after an alert fires. A well-designed escalation path has three properties: it routes the alert to the person with the most relevant expertise for initial triage, and it has clear ownership. Hence, there is never ambiguity about who is responsible for the next action, and it has defined timeboxes so alerts do not sit unacknowledged.

For AI product incidents, the typical escalation path starts with the on-call engineer, who triages whether the issue is infrastructure-related (latency, pipeline failure) or model-quality related. Infrastructure issues follow standard software incident protocols. Model quality issues escalate to the ML engineer or data scientist on call, who determines the scope and severity and decides on an immediate mitigation path. The product owner is notified of all critical and emergency incidents and is responsible for stakeholder communication and go/no-go decisions on fallback or rollback. Executives are looped in when the incident has been confirmed to have a customer-facing impact that will require external communication.

8.6 The Incident Runbook for Model Failures

A runbook is a documented, step-by-step protocol for responding to a class of incidents. Having a runbook does not mean every incident will be neat and predictable. It means your team does not have to invent a response process from scratch at 2 a.m. while also managing a live customer impact. The cognitive overhead savings alone justify the upfront investment in documentation.

The model failure runbook has five phases: Alert, Triage, Contain, Resolve, and Review.

8.6.1 Phase 1: Alert

An automated alert is triggered by the monitoring system. The on-call engineer acknowledges within the SLA defined for that severity level—fifteen minutes for critical, five minutes for emergency. The engineer opens the incident channel, posts the alert details, and begins the Triage phase. The product owner is notified immediately for critical or emergency severity.

8.6.2 Phase 2: Triage

Triage answers the question: What kind of problem is this? The on-call engineer checks the four monitoring layers in sequence. Start with operational metrics: is there a latency spike, elevated error rate, or pipeline failure? If yes, this is an infrastructure incident, not a model quality incident, and follows the infrastructure runbook. If operational metrics are healthy, check data quality: are upstream features arriving as expected, with distributions consistent with recent baselines? If not, a data pipeline issue is likely the cause. If data quality is healthy, check model performance metrics: has accuracy, calibration, or the primary proxy business metric moved outside the alert threshold? If yes, this is a model quality incident. If all three layers are healthy but business metrics have still declined, the issue may be external—a competitor action, a campaign change, or seasonality—and requires business investigation rather than a model incident response.

Diagram 8.5 – Incident Triage Decision Tree

Diagram 8.5 – Incident Triage Decision Tree

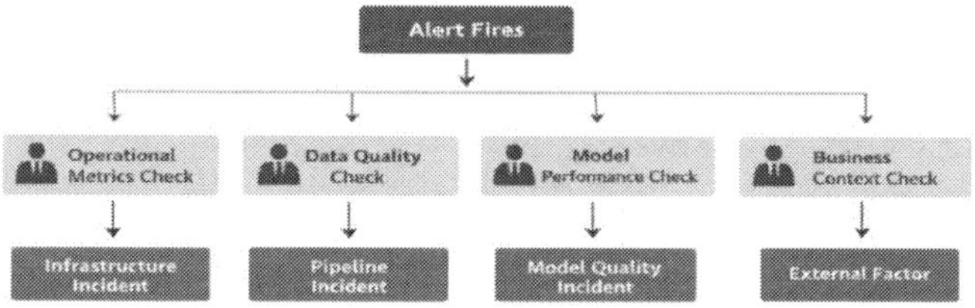

8.6.3 Phase 3: Contain

Containment stops the bleeding while a permanent fix is prepared. The primary containment options for model failures are: activate a fallback model (a prior stable version or a simpler rule-based system if one exists), adjust decision thresholds to trade precision for recall or vice versa depending on which error type is less harmful in the current situation, throttle traffic to the affected model endpoint, or turn off the AI-powered feature and serve a static or rule-based alternative. The choice of containment action depends on which option can be executed fastest with the lowest risk of making the situation worse. Document the containment action and its business impact (estimated percentage of users affected, change in user experience) in the incident channel.

8.6.4 Phase 4: Resolve

Resolution is the permanent fix. For model quality incidents, resolution options include: threshold recalibration (hours), targeted retraining on recent data (hours to days), full retraining and re-evaluation (days), or model replacement (days to weeks). The ML engineer recommends the resolution path based on root cause analysis from the triage phase. The product owner approves the resolution path after weighing the time cost against the business impact of the degraded or contained state. The resolution goes through the standard staged rollout process—not a direct push to production, even under time pressure. Post-rollout, monitoring confirms that performance metrics return to baseline before the incident is closed.

8.6.5 Phase 5: Review

Every incident above warning severity gets a postmortem within five business days. The postmortem is blameless, brief, and action-oriented. It documents the timeline, the root cause, the business impact (estimated), the containment and resolution steps, and the preventive actions that will reduce the likelihood or impact of recurrence. Preventive actions are added to the backlog with owners and due dates. The product owner is responsible for ensuring that postmortem actions are prioritized appropriately, given their risk-reduction value relative to other backlog items.

8.7 Retraining Triggers and Decision Points

The question of when to retrain is one of the most consequential decisions in the AI product ownership role, and it is rarely made with complete information. Retraining consumes compute, engineering time, and evaluation resources. It also introduces risk—a new model version is a new source of potential issues. But failing to retrain when the model has drifted consumes user trust and business value at a rate that usually far exceeds the cost of retraining. The decision framework needs to make this trade-off explicit.

8.7.1 Scheduled vs. Triggered Retraining

There are two basic retraining philosophies: scheduled and triggered. Scheduled retraining runs on a fixed cadence regardless of observed model performance. Triggered retraining runs when one or more monitoring signals cross a defined threshold. Most mature products use a hybrid: a scheduled cadence as a backstop (ensuring the model is never more than a certain age, regardless of performance) combined with triggered retraining for detected drift, which requires a faster response.

The right scheduled cadence depends on how fast your domain evolves. E-commerce recommendations may need weekly retraining. Fraud detection may be needed daily. Industrial anomaly detection on stable equipment may be needed monthly. Set your scheduled cadence based on the drift velocity you observed during your evaluation phase and adjust it over time as you accumulate production data.

8.7.2 The Retrain vs. Patch vs. Replace Decision

When a trigger fires, you choose between three response types. Retrain: use the existing model architecture and feature set, but update the weights with newer data. This is the default response for drift caused by a distributional shift in a well-functioning model. Patch: adjust hyperparameters, decision thresholds, or feature preprocessing without retraining the underlying model. This is appropriate when the root cause is calibration drift or a correctable systematic error rather than genuine model degradation. Replace the existing model with a fundamentally different approach. This is warranted when the root cause analysis reveals that the model architecture or feature set is no longer appropriate for the problem, when drift is so severe that retraining cannot recover performance, or when a new model type is available that offers a step-change in quality.

The replacement decision is the most expensive and should be made deliberately, not under the pressure of an active incident. If you find yourself considering replacement during an incident, contain first with a rollback or fallback, then make the replacement decision in a calmer setting with full evaluation data.

8.7.3 Data Governance for Retraining

Retraining requires fresh, properly labeled training data that is compliant with your data governance policies. This sounds obvious, but it is a common stumbling block: teams discover during an incident that they do not have recent labeled data available for retraining because the labeling pipeline was not maintained, the ground-truth signal has a long lag, or the data has not undergone the required privacy review. Establish a continuous data flywheel before you need it. Define the ongoing labeling, sampling, and compliance workflow as a product requirement rather than an afterthought.

8.8 Tooling Choices and Monitoring Investment Priorities

The MLOps tooling market is large, noisy, and evolving rapidly. Product owners are frequently asked to make tool selection decisions without a clear framework for distinguishing the essential from the aspirational. This section provides that framework.

8.8.1 The Build vs. Buy Decision

The default recommendation for most organizations is to buy or use open-source tooling rather than build custom monitoring infrastructure. The marginal cost of custom monitoring infrastructure is almost always underestimated. It requires ongoing maintenance, it diverts engineering capacity from model improvement, and it rarely achieves the robustness and feature richness of dedicated platforms. Reserve custom development for the cases where a standard tool genuinely cannot meet a requirement—usually a domain-specific metric calculation, a compliance constraint, or an integration with a legacy system that no platform supports.

Cloud-native options (AWS SageMaker Model Monitor, Google Vertex AI Model Monitoring, Azure Machine Learning's model monitoring features) are the lowest-friction starting point for teams already in those cloud environments. They handle infrastructure, storage, and many standard statistical tests out-of-the-box. Purpose-built MLOps platforms (Evidently AI, Arize AI, WhyLabs, Fiddler AI, among others) offer deeper monitoring capabilities, explainability features, and better observability dashboards at the cost of an additional vendor relationship. Open-source tools (Prometheus, Grafana, Great Expectations, Evidently open-source) are appropriate for teams with strong engineering capacity who want full control and low licensing cost.

8.8.2 Investment Prioritization

If you are building a monitoring capability from scratch and need to sequence the investment, prioritize in this order. First: operational monitoring (latency, errors, costs). This is table stakes and is usually achievable with existing infrastructure tooling already in your stack. Second: data quality monitoring on the most critical features. Instrument the top five to ten features by model importance weight—third: proxy business metric tracking tied directly to the model's outputs. If you can only have one model-quality signal and the ground truth is delayed, the proxy business metric is often the most actionable. Fourth: statistical drift detection on feature distributions. Fifth: full ground truth performance evaluation. Sixth: explanation and attribution monitoring for regulated or high-stakes use cases.

You do not need to complete step six before going to production. But you should have steps one through three in place before launch, and steps four and five within the first ninety days. The absence of steps four and five is what leads to the scenario in the opening of this chapter.

8.9 The Observability Dashboard

An observability dashboard is the shared artifact that makes monitoring social. It is the place where the product owner, the ML engineer, the data scientist, and the business stakeholder can all see the same picture of model health at the same time. A good dashboard builds shared situational awareness. A poor dashboard—overloaded with metrics, lacking context, or too technical for non-engineers—creates the illusion of visibility without the substance.

8.9.1 What to Show

The top section of the dashboard shows the business outcome metrics: conversion rate, CSAT, task completion rate, or whatever the primary business KPI is. These should be displayed with trend lines showing the recent trajectory and comparison to the prior period. Business stakeholders look here first. If business metrics are healthy, the conversation stops. If they are declining, the rest of the dashboard explains why.

The second section shows model performance metrics: your primary accuracy or quality metric, calibration error, and the confidence distribution of recent predictions. Trend lines with alert threshold bands make it easy to see at a glance whether performance is inside or outside acceptable bounds.

The third section shows data quality signals: upstream feature health, pipeline success rates, and any active data quality alerts. The fourth section shows operational metrics: p50 and p99 latency, error rate, throughput, and cost per inference. These four sections, kept clean and uncluttered, give every stakeholder the information they need for their role without requiring them to navigate a sprawling metrics catalog.

8.9.2 Who Uses It and How Often

Design your dashboard with three distinct user types in mind. The product owner reviews it weekly in the operational rhythm meeting and references it during any incident. The ML engineer or data scientist uses it daily as a professional hygiene check and uses the drill-down capabilities to investigate anomalies. Business stakeholders access it monthly to confirm that the AI product is contributing to its intended outcomes. On-call engineers need the operational section at a glance, with easy access to the model performance section when an alert fires.

The weekly operational rhythm is the mechanism that keeps monitoring from being passive. Reserve fifteen to twenty minutes in the weekly team sync to walk through the dashboard together. When everything is green, the review takes five minutes and reinforces confidence. When something is amber or red, the review becomes the triage session. This habit prevents the pattern of discovering drift only when a customer complains.

8.9.3 Dashboard Design Principles

Keep the top-level view to a single screen with no scrolling. Every metric that requires scrolling to find will not be checked under time pressure. Use color coding consistently and sparingly—green for within threshold, amber for warning, red for critical—and make sure the color coding is calibrated to your actual alert thresholds so that visual status matches alert status. Include a "model health summary" indicator at the top of the page: a single green/amber/red status that aggregates the overall health signal. This is the metric you share in the executive status update. Beneath it, every other section explains what drives that summary.

8.10 Vignette: Early Drift Detection Averts a Major Customer Impact

Priya was the product owner for a B2B software company's customer health scoring model—the system that ranked accounts by churn risk and prioritized them for the customer success team. The model had been running in production for eight months with stable performance. Then, over the course of three weeks in Q3, something shifted.

Priya's team had set up weekly drift monitoring using the Population Stability Index on the model's top twelve input features. During the third week, the PSI for one feature—product usage frequency in the trailing fourteen days—crossed 0.2, triggering a warning alert. The feature had historically been stable. The spike corresponded to a product update that changed how the engineering team logged usage events—a measurement change, not a behavior change.

The ML engineer traced the alert within a day. The feature's distribution had shifted because the instrumentation had changed, not because customer behavior had changed. Left undetected, this would have caused the model to systematically mis-score accounts: customers with healthy usage would have appeared riskier, and the customer success team would have been misallocating attention based on corrupt inputs. Priya escalated to the product engineering team to correct the logging, and in parallel, initiated a targeted feature normalization patch that could be applied while the logging fix was deployed.

The full containment-to-resolution cycle took six days. A rough estimate of accounts that would have been mis-scored over a full quarter—had the drift not been caught—was 300, with a customer success team misallocation that the commercial team estimated at 4 to 6 missed renewal interventions. No customers experienced degraded service. The customer success team noticed nothing amiss. The incident postmortem produced one action: add automated regression tests for instrumentation changes that affect monitored features.

The investment that made the outcome possible was modest: a drift-monitoring configuration that took one ML engineer three days to set up, a documented alert-response protocol, and a weekly review habit that meant Priya saw the amber alert within 18 hours of it firing, rather than discovering it through a customer complaint.

8.11 Manager's Checklist

Use the following checklist to determine whether your AI product is production-ready from a monitoring standpoint. Revisit it quarterly.

- Operational monitoring is in place and tested: latency, error rates, throughput, cost per inference.

- Data quality monitoring covers the top ten features by model importance, with baseline distributions captured at launch.
- At least one proxy business metric is being tracked in near-real-time and tied directly to model outputs.
- Alert thresholds are set for at least warning and critical severity levels, with documented rationale for each threshold.
- Escalation paths are defined and documented: who receives each alert type, with contact information and SLAs.
- Ground truth collection pipeline is operational, with a documented plan for handling ground truth lag.
- Statistical drift detection is configured for both data drift and performance degradation.
- The incident runbook has been reviewed and tested by the on-call team in a tabletop exercise.
- Rollback and fallback mechanisms exist and have been tested in a non-production environment.
- All key stakeholders have reviewed the observability dashboard, and it is accessible without special permissions.
- Retraining triggers are documented: scheduled cadence, drift-based triggers, and the retrain/patch/replace decision matrix.
- A continuous data flywheel is in place to ensure fresh labeled data is available for retraining when needed.
- Postmortem process is documented and has been socialized with the team.

- Monitoring investment has been sequenced by priority, with a roadmap for reaching full coverage within ninety days of launch.
- Weekly dashboard review is on the team's recurring calendar as a standing agenda item.

8.12 Takeaway

Monitoring is the mechanism that converts a model deployment into a maintained product. Without it, you are not operating an AI product—you are operating a ticking clock. The model will drift. The data will shift. The world will change. These are not failure conditions to be ashamed of; they are inherent properties of deploying a learned system in a dynamic environment. The product owner's job is not to prevent drift—it is to detect it early enough that the response is planned, contained, and invisible to the user.

The minimal monitoring stack is achievable. Four signal categories, two alert severity levels, a documented runbook, a clear escalation path, and a weekly review habit are enough to transform your team's operational posture from reactive to proactive. You do not need a sophisticated MLOps platform on day one. You need the discipline to treat monitoring as a first-class product requirement rather than an infrastructure afterthought.

The manager's lens on monitoring is simple: if you cannot see the model, you cannot manage the model. Observability is the foundation of all the other accountabilities in this guide—evaluation, release, human oversight, and stewardship. Invest in it early, maintain it consistently, and your AI product will earn user trust through reliability rather than losing it through silent degradation.

9 Release Management, Sunsetting, and Long-term Stewardship

Opening Scenario:

The quarterly infrastructure review lands on Elena's desk on a Tuesday afternoon. She is the AI product owner for her company's analytics platform, and three items in the report stop her in her tracks. The first: a legacy sentiment-analysis model provisioned two years ago is consuming thirty percent of the team's machine-learning infrastructure budget. The second: that model feeds exactly four internal dashboards, used by eleven people. The third: there is no documentation explaining why the model was built this way, who owns it, or what would break if it were turned off.

Elena's situation is not unusual. Across enterprises, AI products accumulate quietly. Models that once served a strategic purpose linger long after the need has passed. Infrastructure costs accrue. Maintenance burden grows. The team that built the model has moved on, and the institutional knowledge has evaporated. When the decision to sunset finally arrives—and it always does—it arrives late, under pressure, without a plan.

This chapter is about preventing that outcome. It is a dual playbook: how to release AI products with the discipline that prevents technical debt from forming in the first place, and how to sunset them with the strategic intentionality that recovers resources and protects stakeholder relationships. Both disciplines are sides of the same coin. Without release rigor, models drift into unmanaged states. Without sunset planning, those unmanaged states become permanent fixtures.

9.1 Why It Matters

Release management and sunsetting are not administrative housekeeping tasks. They are strategic decisions that determine whether your AI portfolio grows in health or in entropy. Every model you deploy without a clear release protocol is a future incident waiting to happen. Every model you fail to sunset on schedule is a direct tax on the resources available for higher-value work.

AI products are different from traditional software in a critical way: they degrade silently. A feature toggle left on too long causes configuration drift. A model left running without retraining tends to drift away from real-world data distributions. A rollout without proper user segmentation produces unpredictable variance in user experience. The compound effect of these small failures is an AI portfolio that becomes progressively harder and more expensive to manage.

Managers who build release discipline into their operating rhythms stop problems before they cascade. Managers who build sunset discipline into their portfolio reviews recover resources that would otherwise be silently consumed. Both disciplines require the same underlying capability: the ability to make structured decisions under uncertainty, communicate them clearly to stakeholders, and execute them in a way that protects users and preserves institutional knowledge.

9.2 Release Management Practices for AI Products

Software release management has a long and well-documented history. Branching strategies, continuous integration pipelines, deployment gates, and rollback protocols are mature practices. AI product release management borrows from this tradition but introduces challenges that traditional software does not face.

The core difference is that AI model behavior is probabilistic rather than deterministic. A traditional software deployment either works or it does not. An AI model deployment works on average, across a distribution of inputs, subject to assumptions about the real-world data it will encounter. That probabilistic character demands a different release philosophy: one that prioritizes measured exposure, observable outcomes, and fast feedback loops over binary go/no-go decisions.

9.2.1 What Makes AI Releases Different

Three properties of AI products distinguish their release requirements from those of traditional software.

First, performance is population-dependent. A model that achieves ninety-two percent accuracy on your holdout dataset may perform at eighty-five percent on the first wave of production users if their request distribution differs from your training data. You cannot know this before release. You can only design the release to surface it quickly and limit its impact.

Second, failure modes are often silent. A traditional software bug produces an error message, a crash, or a failed transaction. An AI failure often produces a confident but wrong answer. Users may not report it. Monitoring may not catch it immediately. The first signal is frequently a downstream business metric—customer satisfaction, task completion rate, escalation volume—that moves in the wrong direction weeks after the causative release.

Third, rollback is more complex. Rolling back traditional software means redeploying a previous artifact. Rolling back a model means choosing which version to revert to, ensuring the serving infrastructure supports that version, and deciding whether the data processed during the failed deployment needs to be reprocessed. The answer to that last question is often yes, which makes rollback a non-trivial operational event.

9.2.2 The Release Protocol

A release protocol for an AI product should define the following elements before any deployment begins: the target metric thresholds that constitute success, the sentinel metrics that constitute a rollback trigger, the user population that will receive the new model and in what proportion, the observation period before expanding the rollout, and the communication plan for each stage of the release.

These elements should be documented in a release brief—a one-to-two-page document that any stakeholder can read to understand what is being deployed, to whom, under what conditions, and with what safety mechanisms. The release brief is the accountability artifact. It forces the team to make explicit decisions before the pressure of a live deployment forces implicit ones.

Diagram 9.1 – AI Release Protocol Overview

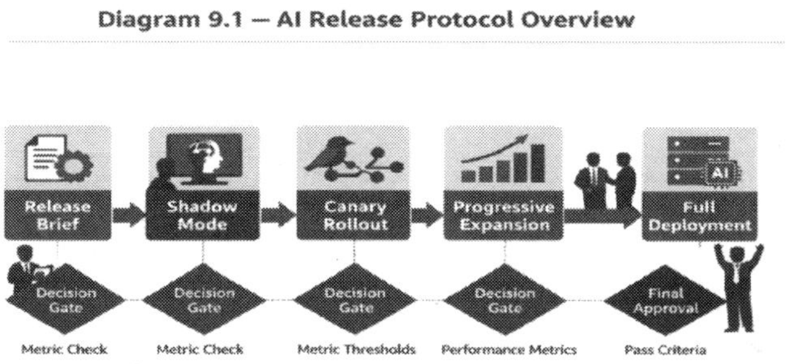

9.3 Feature Flags and User Segmentation for AI Rollouts

Feature flags are the primary mechanism for controlling the exposure of AI models in production. A feature flag is a configuration parameter that determines which version of a model a given user or request receives, independently of the deployment artifact itself. Flags allow you to deploy a new model to production infrastructure while exposing it only to a defined segment of traffic.

For AI products, feature flags serve two purposes that go beyond their traditional use in software. First, they enable controlled exposure to manage the probabilistic risk described above. Second, they create the infrastructure for structured experimentation: the ability to measure model performance as a randomized trial, with a control group receiving the existing model and a treatment group receiving the new one.

9.3.1 Designing Your Segmentation Strategy

User segmentation for AI rollouts should be driven by risk profile, not by convenience. The instinct is to start with the most enthusiastic users—internal teams, beta users, early adopters. This instinct is partially correct but needs to be balanced against the requirement that your early exposure group be representative of the broader user population in the ways that matter most for model performance.

A segmentation strategy should answer three questions. First, which users have the highest tolerance for failure? Internal users and beta participants typically qualify. Second, which user behaviors are most likely to surface edge-case failures in the new model? Users with atypical request patterns, languages, or use cases should be included early to expose failures before they reach the mainstream population. Third, which users would cause the most business damage if they received a poor experience? These users should be among the last to receive the new model.

The practical output of this analysis is a rollout ladder: a sequence of user segments, each larger than the last, with defined observation periods and success criteria at each rung. A typical ladder for a medium-complexity AI feature might look like this: internal team (0.5 percent of requests), opt-in beta users (5 percent), lower-risk segment of production users (20 percent), broader production (60 percent), full rollout (100 percent). Each rung requires a hold time—typically three to seven days—and a metric review before proceeding.

9.3.2 Instrumentation Requirements for Flag-Based Rollouts

Feature flags only deliver value if the observability infrastructure can report performance by flag variant. Before enabling a flag-based rollout, confirm that your monitoring stack can segment every key metric by model version. This means request-level logging that includes the flag variant identifier, aggregation that can produce variant-level dashboards, and alerting that can trigger on variant-specific threshold violations. A rollout where you cannot distinguish the new model's performance from the old model's performance is not a controlled rollout—it is a full deployment with extra steps.

Diagram 9.2 – Feature Flag Rollout Ladder with Segmentation

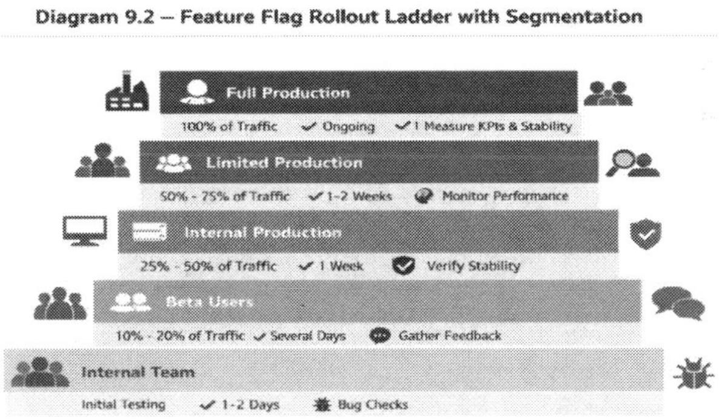

9.4 The Rollback Playbook

A rollback playbook is the operational document that defines what happens when a deployed model needs to be reversed. It answers three questions: when to roll back, how to roll back, and how to communicate the rollback. The playbook should be written before the release, reviewed by all parties, and stored in a location that is accessible during an incident without requiring system access.

9.4.1 Rollback Criteria

Rollback criteria are the specific, measurable conditions that trigger an immediate return to the previous model version. They are distinct from the success thresholds in your release brief. Success thresholds define the conditions for expanding a rollout. Rollback criteria define the conditions under which it should stop.

Effective rollback criteria share three properties. They are quantitative: not "accuracy seems lower" but "task completion rate falls below eighty percent for three consecutive measurement periods." They are time-bounded: the measurement period is defined in advance, so that a single bad hour does not trigger an unnecessary rollback, nor does sustained degradation go unaddressed. At the same time, the team debates whether to act. And they are pre-committed: the team agrees to the criteria before the deployment, not after the metrics start moving, when cognitive biases and sunk-cost fallacies can distort judgment.

Common rollback criteria categories for AI products include: primary model performance metric falling below a defined floor; error rate exceeding a defined ceiling; latency at the ninety-fifth percentile exceeding a defined threshold; downstream business metrics—such as escalation rate, abandonment rate, or customer satisfaction score—moving adversely by a defined magnitude; and any safety or compliance violation, regardless of magnitude.

9.4.2 **Rollback Process**

The rollback process should be executable by any member of the on-call rotation, not only by the ML engineer who built the model. This means the process must be documented step by step, tested in a non-production environment before the release, and owned by a role rather than a person. The steps typically include: decision to roll back (made by the on-call lead), notification to the incident channel, execution of the rollback (feature flag update or model version revert, depending on the deployment architecture), verification that the previous model version is serving correctly, and a hold period to confirm that metrics are recovering.

One common failure in rollback execution is the assumption that rolling back the model is sufficient to restore the previous user experience. In many cases, it is not. If the new model was serving recommendations, routing decisions, or content classifications that have already been acted upon by users or downstream systems, a model rollback does not undo those actions. The rollback playbook should identify the class of model decisions that are reversible and those that are not, and define the remediation approach for the latter category.

9.4.3 Rollback Communication

Communication during a rollback follows the same principles as communication during any operational incident: fast, factual, and proportionate. The first communication should go out within fifteen minutes of the rollback decision to the stakeholders most affected by the deployment. It should state what happened, what the team is doing, and when the next update will arrive. It should not speculate on root cause or commit to a timeline for the new model's return to production.

A post-rollback review should be scheduled within forty-eight hours. The review should produce a written document that captures: the timeline of events, the specific metric behavior that triggered the rollback, the root cause or most probable hypothesis, the changes required before redeployment, and the process improvements that would detect the problem earlier in the next release cycle. This document is not a punishment artifact. It is an institutional memory asset that prevents the same failure in future releases.

9.5 Sunset Strategy and the Sunset Decision Matrix

Sunsetting an AI product is a strategic decision, not an operational one. It requires the same deliberate analysis as a build-versus-buy decision or a major infrastructure investment. The instinct to leave a working model running—

even one that is underperforming or over-consuming resources—is powerful. It feels safer than the alternative. The model is known. The failure modes are understood. The users have adapted to its quirks. Change introduces risk.

This instinct is often wrong. The real risk is not the risk of change. It is the compounding cost of inaction: infrastructure spend that grows as the model ages, maintenance burden that accumulates as the codebase drifts from current standards, and opportunity cost measured in the higher-value work that the resources could be funding. A manager who fails to sunset a legacy model on time is not managing risk. She is deferring it at interest.

9.5.1 The Sunset Decision Matrix

The sunset decision matrix is a structured tool for evaluating whether a model should be retired, rebuilt, or maintained. It evaluates four dimensions: strategic value, operational cost, technical health, and migration feasibility.

Strategic value measures the model's contribution to current business objectives. A model that was central to a product strategy two years ago may be marginal to the current strategy. The question is not whether the model does something useful, but whether what it does is still aligned with where the organization is investing. A model that serves four dashboards consumed by eleven users, when the organization has shifted its analytics strategy to a new platform, has low strategic value regardless of its technical performance.

Operational cost measures the total cost of keeping the model running: compute cost, engineering maintenance time, incident response burden, and the opportunity cost of those resources. Operational cost should be calculated on a per-unit-of-value basis, not in absolute terms. A model that costs fifty thousand dollars per year to maintain but generates ten million dollars in measurable business value has a favorable cost ratio. A model that costs $20,000 per year to maintain but is used by 11 people for low-stakes dashboard views does not.

Technical health measures the condition of the model's codebase, data pipelines, and supporting infrastructure. A model built on deprecated frameworks, trained on data that is no longer being refreshed, and integrated with systems that have since been replaced has poor technical health. Poor technical health is a leading indicator of future incidents and a multiplier on maintenance costs.

Migration feasibility measures how difficult it would be to move the model's users and use cases to an alternative. If a direct replacement exists, migration feasibility is high. If the use case is unique and a replacement would require significant development investment, the feasibility is lower. Migration feasibility does not determine whether to sunset—a high-cost, low-value model should be sunset regardless—but it does determine the timeline and the required investment in migration support.

Diagram 9.3 – Sunset Decision Matrix: Four-Quadrant Evaluation

Diagram 9.3 – Sunset Decision Matrix: Four-Quadrant Evaluation

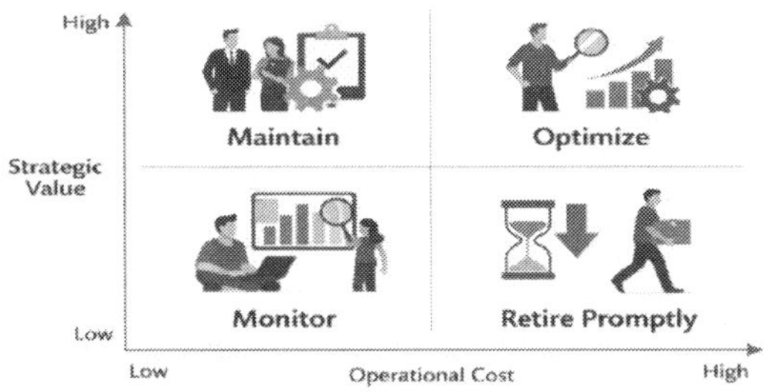

9.5.2 Rebuild vs. Retire

When the sunset decision matrix points toward retirement, the next question is whether to rebuild the capability or retire it entirely. Rebuilding is appropriate when the strategic value is high, but the current implementation is technically unhealthy. Retirement is appropriate when the strategic value is low, regardless of technical health.

Two cognitive biases frequently distort the rebuild-versus-retire decision. The first is the sunk-cost fallacy: the team has invested significant effort in the existing model, and retirement feels like a waste. The solution is to evaluate the decision based only on forward-looking costs and benefits. Past investment is not recoverable regardless of the choice made today. The second bias is feature attachment: stakeholders who use the model have built workflows around it, and they resist retirement even when a better alternative exists. The solution is to involve these stakeholders in the migration planning process early, so they have agency over the transition rather than it being imposed on them.

9.5.3 Cost-of-Maintenance Analysis

Before presenting a sunset recommendation, build a cost-of-maintenance analysis. This document quantifies the total cost of keeping the model running for one, two, and three additional years, including infrastructure cost at current utilization, engineering time at current maintenance burden, projected cost increases as the model ages and its dependencies become harder to maintain, and the opportunity cost of the compute and engineering resources at their next-best use. Present this analysis alongside the migration cost and timeline for the recommended alternative. The comparison makes the economic case for the sunset decision visible to stakeholders who might otherwise focus only on the risk of change.

9.6 Migration Planning for Model Transitions

Migration planning is the operational discipline that makes a sunset decision executable. A decision to sunset a model without a migration plan is not a decision. It is a declaration of intent with no path to action. Migration planning defines how users, use cases, and data flows move from the old model to the new one, and how that transition is managed to preserve user trust and business continuity.

9.6.1 Mapping the Dependency Graph

The first step in migration planning is mapping every dependency on the legacy model. This includes direct consumers—the applications, dashboards, and workflows that call the model API directly—and indirect consumers—the downstream systems that receive the model's outputs through intermediate layers. Dependency mapping is often more complex than it appears. Models that have been running for two or more years often have consumers built without the original team's knowledge, as other teams have found ways to incorporate the model's outputs into their own workflows.

The dependency map should be produced by inspecting call logs and API metrics, not by asking stakeholders to self-report. Self-reporting undercounts dependencies consistently. Teams forget about integrations they built, or do not recognize that a data pipeline they rely on ultimately traces back to the legacy model. A systematic audit of API call logs, data lineage records, and integration documentation will produce a more complete picture.

9.6.2 Defining Migration Windows

A migration window is the period during which a specific consumer or use case transitions from the legacy model to the replacement. Migration windows should be defined for each consumer in the dependency map, sequenced from lowest to highest risk, and scheduled to avoid business-critical periods—end-of-quarter reporting cycles, peak usage periods, and planned system maintenance windows.

Each migration window should include a parallel-running period, during which both the legacy model and the replacement model are active, with outputs compared programmatically. The parallel run allows the team to confirm that the replacement model's behavior is acceptable before the legacy model is turned off. The length of the parallel run depends on the use case: high-stakes decisions may require weeks of parallel operation, while low-stakes dashboard metrics may require only days.

9.6.3 Data Retention During Model Transitions

Every model transition raises a data retention question: what happens to the data generated by or used to train the legacy model? The answer has legal, operational, and institutional-memory dimensions that must be resolved before the legacy model is decommissioned.

From a legal perspective, data generated by the model—predictions, classifications, recommendations—may be subject to retention requirements imposed by regulation, contractual obligation, or audit requirements. These requirements must be reviewed and satisfied before decommissioning.

From an operational perspective, the training data and feature-engineering logic from the legacy model are valuable inputs for the replacement model's development. They should be archived in a queryable format, not discarded.

From an institutional-memory perspective, the model's performance history—its metrics over time, the incidents it experienced, the retraining events it underwent—is a documented record of what the organization learned about this problem domain. That record should be preserved in a format that a future team can access when building on the same problem space.

Diagram 9.4 – Migration Window Sequencing and Parallel-Run Architecture

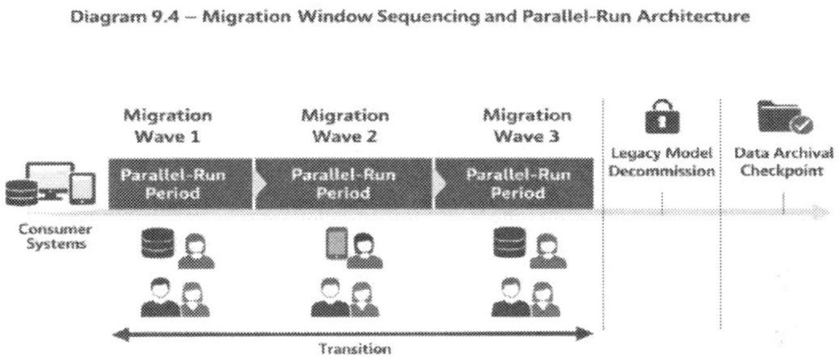

9.7 Stakeholder Communication During Sunsets

Stakeholder communication is the most frequently underinvested element of a sunset program. Teams focus on the technical migration and treat communication as a secondary activity, sending a brief announcement shortly before the legacy model is turned off. This approach produces the worst possible outcome: stakeholders who feel blindsided, who have not had time to adapt their workflows, and who will resist future model transitions.

Effective sunset communication follows a layered approach that begins far earlier than most teams expect. The announcement of the sunset decision should happen at the moment the decision is made, not at the moment the migration begins. Users and downstream teams need time to understand the implications, raise concerns, and prepare their own adaptation plans. Giving them that time is not a courtesy. It is a risk management strategy that reduces the probability of last-minute escalations and missed migration windows.

9.7.1 Communication Templates

A sunset communication program typically requires three message categories.

Decision announcement: Sent at the moment the sunset decision is made. Content: what is being sunset, why, when the legacy model will be decommissioned, what the replacement is, and how stakeholders can raise questions or concerns. Tone: factual, forward-looking, acknowledging the disruption while explaining the strategic rationale. Audience: all direct consumers of the legacy model, and any stakeholders who may be affected by downstream changes.

Migration readiness update: Sent at regular intervals during the migration window, typically monthly. Content: migration progress by consumer segment, current parallel-run status, any changes to the decommission timeline, and the specific actions required from each stakeholder group in the upcoming period. Tone: operational, specific, action-oriented. Audience: technical owners of consumer systems and operational managers of affected workflows.

Decommission notice: Sent two to four weeks before the legacy model is turned off. Content: the specific date and time of decommissioning, the verification steps consumers should take to confirm they have migrated, the escalation path for any consumer that cannot complete migration by the deadline, and the data archival plan for legacy model outputs. Tone: firm on timeline, supportive on execution. Audience: all consumers and their management chains.

9.7.2 Managing Resistance

Stakeholder resistance to model sunsets is predictable and manageable. The most common forms of resistance are: claims that the legacy model produces outputs that the replacement cannot replicate, concerns about the effort required to update integrations, and general attachment to the known system over the unknown one.

The most effective response to all three forms of resistance is the same: data. Parallel-run results that demonstrate the replacement model's output quality directly address the performance concern. Dependency mapping and migration support directly address the concern about integration effort. User testing with the replacement model, conducted with resistant stakeholders before the migration window, directly addresses the attachment concern. Resistance that persists after these interventions should be escalated through the stakeholder's management chain, with the cost-of-maintenance analysis as the supporting evidence.

9.8 Knowledge Capture for Successor Products

Every sunset model carries institutional knowledge at risk of being lost. The team that built it understands why certain design choices were made, which edge cases proved most challenging, which user behaviors the model handles poorly, and which data quality issues have never been fully resolved. If that knowledge is not captured before the model is decommissioned, it disappears—and the team building the successor model will spend months rediscovering what the previous team already knew.

Knowledge capture is not a documentation exercise in the traditional sense. It is a structured debriefing process that extracts tacit knowledge from the people who built and maintained the model, and converts it into explicit artifacts that a future team can use. The process should be completed before the migration window closes, while the original team members are still available and their memories of the model's behavior are still fresh.

9.8.1 The Knowledge Capture Artifact Set

A complete knowledge capture for a sunsetted model produces five artifact types.

Model card update: A final version of the model's documentation that includes its intended use case, training data provenance, known limitations, performance characteristics across user segments, and the reason for retirement. This document should be stored in the organization's model registry, linked from the replacement model's documentation.

Failure mode catalog: A documented list of the specific input types, edge cases, and conditions under which the legacy model was known to perform poorly. This catalog is the single most valuable artifact for the successor model's development team, as it defines the test cases and edge-case training data they will need to acquire.

Data lineage record: A complete record of the data sources, transformation logic, and labeling approaches used to train and evaluate the model, including any known quality issues with each source. This record allows the successor team to make informed decisions about which data assets to carry forward and which to replace.

Integration architecture diagram: A diagram of every system that consumed the legacy model's outputs, the format and frequency of those outputs, and the migration approach taken for each consumer. This diagram is the reference document for any future team that needs to understand what the legacy model did and how it connected to the broader system architecture.

Lessons learned document: A narrative account of the most significant challenges encountered during the model's lifetime, the approaches taken to address them, and the conclusions reached about what would be done differently if the model were rebuilt. This document is explicitly subjective and experiential. It captures the team's judgment and intuition in a way that structured documentation cannot.

9.9 The Phased Sunset Playbook

A phased sunset playbook is the operational master document for a sunset program. It brings together the migration plan, the communication calendar, the data retention plan, and the knowledge capture schedule into a single structured timeline. The playbook is owned by the AI product manager and reviewed at the same cadence as other program milestones.

9.9.1 Phase One: Impact Analysis (Weeks One Through Four)

The first phase of the playbook focuses on understanding the full scope of the sunset's impact before any migration work begins. Activities in this phase include: completing the dependency map, building the cost-of-maintenance analysis, identifying the replacement model or confirming that the use case will not be replaced, and conducting initial stakeholder briefings to gather input on migration complexity and timeline constraints.

The output of Phase One is a sunset proposal document that includes the dependency map, the cost-of-maintenance analysis, the proposed migration timeline, and the data retention plan. This document is reviewed and approved by the relevant management stakeholders before migration work begins.

9.9.2 Phase Two: Migration Windows (Weeks Five Through Fourteen)

The second phase executes the migration for each consumer segment in the dependency map, sequenced from lowest-risk to highest-risk. Each migration window includes: parallel-run configuration, metric comparison reports, stakeholder sign-off on parallel-run results, consumer cutover, and post-cutover monitoring. The migration phase is driven by the signed-off consumer list from Phase One. No consumer should be cut over without explicit confirmation that its integration has been tested and its stakeholders have accepted the parallel-run results.

At the end of Phase Two, all consumers should be confirmed to have been migrated. Any exceptions—consumers that have been granted extensions or have not completed migration—should be documented with a specific resolution date and the responsible stakeholder.

9.9.3 Phase Three: Data Retention and Decommission (Weeks Fifteen Through Eighteen)

The third phase executes the decommissioning of the legacy model after all consumers have migrated. Activities include: final data archival in accordance with the retention plan, completion of knowledge capture artifacts, notification of decommission to all stakeholders, teardown of legacy model infrastructure, and a post-sunset review to capture lessons learned for future sunset programs.

The post-sunset review is the most frequently skipped activity in this phase. It should not be. The post-sunset review is the mechanism by which the organization improves its sunset playbook over time. Each sunset program surfaces new complexities, new stakeholder dynamics, and new technical challenges that the playbook did not anticipate. Capturing these learnings and incorporating them into the playbook before the next sunset program begins is how the organization builds the institutional capability to manage AI portfolio transitions at scale.

Diagram 9.5 – Phased Sunset Playbook Timeline

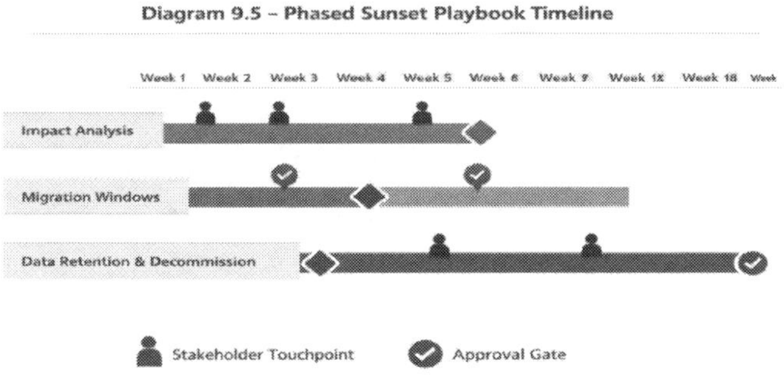

9.10 Vignette: Early Planning Enables a Smooth Migration

Returning to Elena: rather than waiting for the quarterly review to force the decision, she initiates a sunset program the week after receiving the infrastructure report. She spends the first two weeks completing the dependency map—auditing API call logs and interviewing the eleven dashboard users. She discovers two additional consumers of the legacy sentiment model that were not on the original list: a weekly executive summary report and a data pipeline feeding a compliance dashboard. Both had been built by other teams without her knowledge.

With the complete dependency map in hand, Elena builds the cost-of-maintenance analysis. She calculates that the legacy model costs $62,000 per year to run, including compute, engineering maintenance, and incident response. She identifies a replacement model that has already been deployed for a different use case and can be adapted to serve the sentiment analysis function at a total migration cost of twenty-eight thousand dollars. The replacement model will then cost $4,000 per year in compute costs. The payback period for the migration investment is under six months.

She presents the sunset proposal to her management stakeholders in week three. The economic case is clear. Approval comes within the week. She sends the decision announcement to all five consumer teams the same day, eighteen weeks before the planned decommission date. The early notice gives each team enough time to plan their integration work without it becoming an emergency.

The migration proceeds in three waves. Internal dashboards migrate in weeks five and six. The executive summary report migrates in week eight, following a two-week parallel run to confirm output quality. The compliance dashboard migrates in week twelve, the longest parallel run, because the legal team reviews its outputs, and the sign-off process takes longer. By week fourteen, all consumers have migrated. The knowledge capture artifacts are completed in week fifteen, and the legacy model infrastructure is decommissioned in week sixteen.

The resources freed by the decommission—compute budget and one-quarter of an ML engineer's maintenance time—are reallocated to a new model project that has been waiting for capacity. Elena's post-sunset review documents the two hidden consumers as the most significant discovery of the program, and the dependency-mapping protocol is updated to include a mandatory API log audit for all future sunset programs. The next sunset, eighteen months later, runs two weeks faster and encounters no surprises.

9.11 Manager's Checklist

- Write and publish a release brief before every model deployment, including target metrics, rollback criteria, and user segmentation plan.
- Confirm that your monitoring stack can segment all key metrics by model version before enabling any feature flag rollout.
- Define rollback criteria in advance of deployment, with quantitative thresholds and time-bounded measurement periods, and pre-commit the team to honoring them.
- Schedule a post-rollback review within forty-eight hours of any rollback event, and preserve the review document as an institutional memory asset.
- Conduct a quarterly AI portfolio review that evaluates every model against the four dimensions of the sunset decision matrix: strategic value, operational cost, technical health, and migration feasibility.
- Build the cost-of-maintenance analysis for any model that scores poorly on strategic value or operational cost before presenting a sunset recommendation.
- Complete the dependency map using API log audits before beginning any migration planning—do not rely on self-reported consumer lists.
- Send the sunset decision announcement at the moment the decision is made, not at the start of the migration window.

- Define parallel-run periods for each consumer migration window, and require explicit stakeholder sign-off on parallel-run results before cutting over.
- Complete the five knowledge capture artifacts—model card update, failure mode catalog, data lineage record, integration architecture diagram, and lessons learned document—before the legacy model is decommissioned.
- Schedule a post-sunset review within two weeks of decommissioning, and incorporate the findings into the organization's sunset playbook.
- Reallocate freed infrastructure resources to a defined higher-value project within thirty days of decommissioning—do not allow recaptured capacity to be absorbed silently by the existing portfolio.

9.12 Takeaway

Release management and sunsetting are not the glamorous end of AI product ownership. They do not generate the excitement of a new model launch or the satisfaction of a successful pilot graduation. But they are the disciplines that determine whether an AI portfolio remains healthy and strategically aligned over time, or whether it becomes a growing burden of legacy infrastructure, unmaintained models, and accumulated technical debt.

The manager who builds release discipline into every deployment cycle creates the conditions for fast, safe iteration. The manager who builds sunset discipline into every portfolio review creates the conditions for sustained resource availability and strategic flexibility. Both disciplines are operational capabilities that compound over time: each rollout handled well makes the next one easier; each sunset executed cleanly improves the playbook for the next.

The fundamental insight is that AI products are not static artifacts. They are living systems that require active stewardship across their entire lifecycle, from the release brief that defines the conditions for a safe deployment to the decommission notice that closes a chapter and frees resources for what comes next. Managers who accept the stewardship role and build the processes that make it systematic are the ones whose AI portfolios deliver durable value rather than temporary results.

Elena's sentiment-model decision is a model for how that stewardship looks in practice: deliberate analysis, early communication, structured migration, careful knowledge capture, and explicit reallocation of freed resources to higher-value work. That sequence is repeatable. That sequence is a playbook. And a playbook, executed consistently, is how disciplined AI product management becomes an organizational competitive advantage.

10 Conclusion

This book has argued a simple but demanding idea: AI guardrails are not optional safeguards for a few exotic systems; they are the operating system for any organization that wants to use AI at scale and stay trusted. As an IT manager, you sit in the uncomfortable but powerful space between ambition and accountability. You are close enough to the technology to know what it can do, and close enough to the business to feel the risk when things go wrong. The chapters you have just worked through are a playbook for turning that tension into a durable advantage.

We started with intent: why guardrails matter now, and how unmanaged AI quietly creates technical, legal, and reputational debt. We then moved into structure—governance models, roles, and decision rights that give you a clear map of who owns what. From there, we translated principles into repeatable practice: policies that actually show up in workflows, technical guardrails that catch problems before they reach users, human-in-the-loop patterns that keep judgment where it belongs, and risk triage models that focus your scarce capacity where it matters most. Monitoring, drift, and incident response brought an operational lens: not "if something goes wrong" but "when it does, how quickly do we detect, contain, and learn."

The second half of the book focused on building an AI program that can withstand regulatory, organizational, and engineering pressures. You saw how to design compliance once and localize it across jurisdictions, instead of chasing every new rule with a bespoke fix. You embedded guardrails in CI/CD pipelines, model registries, and data workflows so that compliance is enforced by infrastructure, not heroics. Finally, you stepped back to look at culture and maturity: how organizations move from ad hoc controls to an embedded, quietly confident posture where responsible AI is "how we work here."

As you close this book, the most important next step is not to "do everything" but to pick one or two concrete moves and execute them well. That might be creating a simple AI use-case register and risk triage, standing up a lightweight governance working group, or adding fairness and data-quality checks to a single pipeline. Once that works, expand. Guardrails, like any infrastructure, are built layer by layer: policy, process, tooling, culture. If you keep tying each investment back to mission, value, and risk, you'll find it easier to secure sponsorship and sustain momentum over the long term.

AI will only become more capable, more connected to your core operations, and more visible to regulators and the public. Organizations that treat guardrails as friction will spend the next decade reacting—patching incidents, scrambling for audits, and explaining failures. Organizations that treat guardrails as a strategic asset will move faster with more confidence because they know their systems are monitored, explainable, and defensible. Your role is to help your organization join the second group. You now have the frameworks, language, and patterns to do that. The rest is leadership: taking the first step, making the hard calls when trade-offs arise, and building a culture where responsible AI is not a slogan but a habit.

Made in the USA
Monee, IL
12 November 2026

3fef5602-ecb6-4774-801f-635ae4144f83R01